The Monster Trilogy Guidebook

How to Find Bigfoot, Yeti & the Loch Ness Monster

Peter Byrne, F.R.G.S.

Editing and Photograph Coordination
Christopher L. Murphy

hancock
house

ISBN 978-0-88839-723-2

Library and Archives Canada Cataloguing in Publication

Byrne, Peter, 1925–
    The monster trilogy guidebook : how to find Bigfoot, Yeti & the Loch
Ness monster / Peter Byrne ; edited and photograph coordination Christopher L.
Murphy.

    Includes index.
    ISBN 978-0-88839-723-2

    1. Sasquatch. 2. Yeti. 3. Loch Ness monster. I. Murphy, Christopher L.
(Christopher Leo), 1941- II. Title. III. Title: How to find Bigfoot, Yeti & the
Loch Ness monster.

QL89.B97 2012   001.944   C2012-904891-7

Copyediting: Theresa Laviolette
Production: Christopher Murphy, Ingrid Luters
Cover Design: Ingrid Luters

Front Cover Images: Sasquatch, C. Murphy; Yeti, R. Bateman; Loch Ness Monster – Attribution: Stara
Blazkova at the Czech language Wikipedia (Creative Commons provision, refer to Wikipedia for details).
Back Cover Image: P. Byrne

*We acknowledge the financial support of the Government of Canada through the
Canada Book Fund for our publishing activities.*

Printed in South Korea—PACOM

Published simultaneously in Canada and the United States by

HANCOCK HOUSE PUBLISHERS LTD.
19313 Zero Avenue, Surrey, BC Canada V3S 9R9
(604) 538-1114 Fax (604) 538-2262

HANCOCK HOUSE PUBLISHERS
1431 Harrison Avenue, Blaine, WA, USA 98230-5005
(604) 538-1114 Fax (604) 538-2262

Website: **www.hancockhouse.com**
Email: **sales@hancockhouse.com**

DEDICATION

This book is dedicated to all of my companions and fellow adventurers of the years—men and women of courage and zeal—all of whom, often in the face of public disdain and private ridicule, nevertheless joined me and supported me in pursuit of solutions to the last three great mysteries of our planet earth. In doing so, they explored with me—often under extreme weather conditions—in the high Himalaya, carried out strenuous field work in the rugged mountains of the Pacific Northwest, and probed the murky depths of Loch Ness.

They are too many to mention in name here and, indeed, over the course of six decades time has taken its toll to where, sadly, some of them are no longer with us. The names of these who are gone are bright in my memory for their loyalty, courage, and above all, the integrity they applied—always—to their research.

Those who have come after them—the people who work with me today—possess the same great qualities and, like those who have preceded them, apply to their work two things that this kind of phenomenon research demands, two great human attributes: a sense of wonder for the known, and a healthy curiosity for the unknown. I am proud to have all of them as my peers in the great searches and, indeed, honored to be able to dedicate this little work to them.

Contents

Note on Terminology

The words "bigfoot" and "sasquatch" are interchangeable. Generally speaking, "bigfoot" is the American name and "sasquatch" is the Canadian name for the creature. I have chosen not to capitalize either word. I have also chosen to consider the word "sasquatch" and "bigfoot" as both singular and plural terms to avoid the cumbersome or inappropriate terms "sasquatches," "bigfoots" or "bigfeet."

FOREWORD

You'll need more than good boots and a camera (or video recorder) if you want to find and film a "monster"—you'll also need this book. It's refreshing when an iconic individual like Peter Byrne releases his wealth of information. Here, in this book, Peter pulls from his decades of personal experience and shares it with the reader. From the elusive bigfoot in the Pacific Northwest to the mysterious yeti in Nepal, or to Nessie hiding in that deep, cold lake in Scotland, Peter gives the reader not only his colorful history on those monsters, but suggestions on the "how to" in preparing for the trek, and in looking for and finding the creatures.

This detailed guide, written by an expert, is intended to assist those who seek adventure and want to understand wildlife better. It's a "must" book for the outdoor enthusiasts who have the stamina, the patience, and a strong constitution for a challenging undertaking in the wild.

So, if you're going to tromp through the forest of the Pacific Northwest, capture a spectacular view of the Himalayas, or maybe just float around that eerie Loch Ness waiting for your boat to "rise up," this book not only will recommend the best way to get you there, but also will advise you about the equipment you'll need and what to expect. You're sure to have a true adventure in simply reading the book, and will benefit greatly from the experiences of this author.

Peter's knowledge is what led me, and the group of hunters I was involved with, to seek him out in 1971. We eventually met and have been good friends ever since. Our common interest was the bigfoot mystery. Although Peter and I have differences of opinion on how to approach that phenomenon, we both agree that neither of us really know what these creatures actually are—nobody does—and that's the challenge. We do agree, however, that bigfoot is non-violent and should be hunted with only a camera and a good attitude.

— RON MOREHEAD

INTRODUCTION

Our little planet, small and shrinking though it is, still has many mysteries. Among them, three stand out as intriguing, perplexing and challenging. These are: the sasquatch or bigfoot of the Pacific Northwest of the USA and Canada; the yeti, or abominable snow-men, of the Himalaya; and the legendary prehistoric monsters of the deep, cold lake that is known as Loch Ness, in Scotland.

All three phenomena have been known to us for hundreds of years and, at first glance, to many seem to be unfathomable and, as such, impossible to explain and resolve. But the fact is, they can be solved. It's just a question of how to go about doing this and that is what this publication, in three carefully researched sections, tells its readers.

For Americans and Canadians, probably the most intriguing of the trio is the bigfoot mystery, and this is mainly because it exists literally in our own backyards. Ongoing reports of sightings and footprint finds keep it alive, constantly contributing to the growing belief that an extremely primitive hominoid has managed to survive and may well be presently living in the vast forests of the Pacific Northwest.

The yeti of the Himalaya seems to be another primate, smaller and more elusive. Research, much of it conducted by the author, indicates that its habitat is not the eternal snows of the high Hima-layan ranges, but actually the deep, heavily forested valleys that lie below them. In other words, to go after one, an expensive expedition specially equipped with high-altitude equipment is not necessary. Anyone can be involved, and apart from the intrigue of the search, time spent in the glorious terrain of the Nepal Himalaya is always unforgettable.

The Loch Ness monsters present a different challenge. Believed (though not confirmed by scientists) to be remnant, prehistoric ple-siosaurs that survived Scotland's ice age to become landlocked in this ancient loch, these huge creatures with their long necks, snakelike heads and diamond-shaped fins, have been pursued by "Nessie" hunters for years. To date no one has been able to obtain clear, well-defined surface pictures or surface video. But with patience, and the proper approach, this can be done; and, like the bigfoot mystery and the challenge of the yeti, it can be done by anyone with dedication and determination.

The Monster Trilogy examines in detail these last, three great mysteries of planet earth. More than this, it tells you how, with the proper approach, you could be the person to crack any one of them and, in the process, make a spectacular and historic find.

Having said that I wish to make two things clear to the reader. Firstly, I have not personally seen any of the creatures in the trilogy. Nevertheless, I have extensive experience in field work (wilderness expeditions) and with wild animals (which I detail later in this work). These qualifications, coupled with my many years of study in what is termed cryptozoology have, I believe, provided me with insights on resolving animal-related phenomena which will be very valuable to those involved in the search for unrecognized creatures or who wish to get involved in this fascinating field of study. Secondly, with regard to my conclusions on bigfoot or sasquatch, these are based on my own personal investigations of witness reports. In other words, I have not drawn upon the work of others—what I present are all first-hand conclusions based on 125 eyewitness reports (out of many hundreds submitted) that I deemed had very high credibility. This is not to say that the work and conclusions reached by others is not valid, it just means that many things claimed by others were not part of my experience.

—PETER BYRNE
www.petercbyrne.com

PREFACE

This book is the result of requests from many people, young and old, for a sensible and down-to-earth guide book on the great searches that have taken place in the last half century for the bigfoot of the Pacific Northwest, the yeti of the Himalaya, and the mysterious monsters of Loch Ness in Scotland. In other words, a "how to" book that will tell them where to go, what to take in the way of equipment, what to expect in the way of hazards, costs and logistics, and what, with a little bit of luck accompanied by appropriate dedication and tenacity, they may expect to find.

The work is based on the personal research of the author, which in summary includes five expeditions into the Nepal Himalaya, three of them back to back, covering a period of thirty-six months; five exploratory visits to Loch Ness, three of them in the company of members of the esteemed Academy of Applied science of Boston, Massachusett; and more than five decades of personal interest and research in the Pacific Northwest, which period included the design and operation of two multi-million dollar professional, full time research projects (the results of which are still being applied by the author in current research on the bigfoot mystery).

Indicative of the author's background and ability to present to the public an authoritative guide to the phenomena in question is the international recognition he has received for his work in the high Himalaya, at Loch Ness, and in the Pacific Northwest. As a result, today, he has the honor of being a Fellow of the Royal Geographical Society of London, a Member of the Academy of Applied Science of Boston, and a Member Emeritus of the Explorers Club of New York.

Artwork showing a possible likeness of a bigfoot based on what is believed to be an actual photograph (film) of the creature taken in Northern California. Generally, people who see a bigfoot either do not have a camera with them, or are unable to get a photograph for several reasons—the main one being that they did not have enough time. Any known photographs (other than the film mentioned) are too blurry or the subject is too far away to see any meaningful details. *(Artwork/photo: C. Murphy.)*

SECTION ONE The Bigfoot Mystery

CHAPTER 1

Introduction to Bigfoot & Author's Related Background and Experience

So you want to find a bigfoot? You've heard about them, you've read about them, and now you're excited about the mystery and all set to go out and look for one. Well, you're in luck and this is for a few simple reasons:

> One, you've just picked up the definitive guide to bigfoot hunting, written by an expert and reviewed by experts, the book that tells you everything you need to know about how to find your first bigfoot.

> Two, if you live in the USA or in Canada, bigfoot hunting begins, literally, just outside your own back door.

> Three, bigfoot hunting is free. There is absolutely no charge. And whereas research or exploration into, say, the depths of the sea, or outer space, could cost you and arm and a leg, bigfoot hunting costs pennies.

> Four, there are no licenses required, no permits, and no entry fees into the great game of bigfoot hunting. Just a few simple rules and a few simple requirements.

The most important of these requirements, as far as this author is concerned, is that your intentions towards the creatures are peaceful and benign and that you will never, ever, in any way, attempt to harm one of them. Because whatever these marvellous beings may be, there is one thing that we are sure of and that is that they are completely harmless to mankind. For in all of the contact that they have made with people, no one has ever even been threatened by one of them, a fact thoroughly substantiated by more than forty years of professional research on the part of this writer and his associates, some of whom have actually had close encounters with the creatures.

The other requirements are obvious ones and they center around a proper respect for the law in regard to public lands and private property, plus appropriate safety precautions when going into the rugged and sometimes dangerous terrain of the Pacific Northwest. You'll find these all covered in this book. Read it carefully and stay within the law. Also—stay safe.

Lastly, if you feel a little daunted about the prospect of going alone into the great forests of the Northwest to look for one of these creatures, keep in mind that, apart from the author's personal bigfoot projects, there have been no professional, full-time searches for the creatures—no organized scientific field research or scientific expeditions. Which means that you, as an amateur bigfoot hunter, have as much chance as anyone else of finding one, especially if you follow the guidelines contained here.

So, when you set out on your great adventure, remember to take along your copy of *The Monster Trilogy*. It may not lead to your first bigfoot on your very first outing, but it will probably get you closer than anything else presently in print.

In fact, as its author, what I would strongly suggest is, don't leave home without it!

Background and Experience

When someone sets out to write a guide on any subject it is hoped (indeed, it is imperative) that he or she has some background and experience in the topic. So, who is this fellow Byrne and what gives him the right to think of himself as an authority on the bigfoot mystery, with supposed expertise that allows him to offer advice to potential bigfoot hunters? What's his background in the phenomenon? How long has he been involved in it? What was he doing before he became a bigfoot hunter, and what kind of experience does he have that he believes will be useful to others in their research?

All good questions and, as such, deserving of answers, if only so that the reader can feel that the advice that he or she is getting has a good solid background to it, one that includes both qualifications and experience.

This writer entered into the great search for bigfoot in, of all places, a hotel in Kathmandu, Nepal, the old, city center Royal Hotel, now long since defunct. I had just come down from the Himalaya where I had spent a long three years running the American yeti, or abominable snowman, expeditions on behalf of the Southwest Research Institute of San Antonio, Texas. The date was December 31st, the year was 1959 and I and my brother, Bryan, had just arrived at the Royal after a 400-mile hike from eastern Nepal.

Tom Slick *Photo: P. Byrne.*

While we were in the Himalaya, we had mail brought to us once a month by a runner from Kathmandu. Now there was a pile of it waiting for us at the hotel and one of the first things that we opened was a cable from the founder of the Southwest Research Institute, the famous Texan explorer, Tom Slick, welcoming us back from the mountains, congratulating us on concluding three years of high altitude research, and asking us if we would like to take a break from yeti hunting and come to the US and investigate something called the "bigfoot" phenomenon.

I remember my brother looking at me skeptically and laughing. A what? And where? But just a few weeks later I was in Tom Slick's home in San Antonio, pouring over maps of the enormous mountain ranges of the Pacific Northwest—which, with British Columbia, are three times the size of the Nepal Himalaya—and looking at the evidence that he had already collected. The laughing and skepticism had quickly come to an end.

Shortly after that I flew to Monterey, California, picked up a Jeep and a pile of camping gear and drove north to a place called Willow Creek, in Humboldt County, northern California. There I set up the first full-time, professionally run bigfoot research organization, with paid office staff and field researchers. I called the project the Pacific Northwest Bigfoot Project and I ran it for two and a half years until the sad and untimely death of Tom, in an air crash in October 1962, put an end to the operation.

I thereupon went back to Nepal to resume the life that I had been living there when I was first contacted by the San Antonio group, back in 1956. This was as a "white hunter," running professional, big game hunting safaris with rich and sometimes famous clients, in the beautiful forests and grasslands of an area called the White Grass Plains in the country's far southwest.

Going back even before this, which I think is necessary if we plan to look at qualifications here, takes me into the 1940s, when, after four years with the British Royal Air Force in Southeast Asia during World War Two, I worked with an English tea company up in north Bengal. There, in the great forests of the Terai, a belt of land that stretches right across north India all the way to the western reaches of Nepal, my great interest in natural history, one that had started in the Irish countryside as a boy, was kindled and expanded; and this is what I carried with me into twenty years of safari life in the forests of southwest Nepal, took with me into the Himalaya in the searches for the yeti, and then carried forward into the bigfoot field.

In December 1969, after an absence of eight years from the bigfoot scene, I was approached by an old friend of Tom Slick, one C.V. Wood of Los Angeles, and asked if I would be interested in starting up another bigfoot project. I had just spent a year working with the Nepal government to convert my old hunting concession—60,000 acres of magnificent forest and grassland in southwest Nepal—into a protected wildlife sanctuary and, being free of any pressing obligations at the moment, told Wood that I was very interested. The upshot of this was another bigfoot project, the second for me and one that began immediately and lasted for ten years.

For this one I based at first in northern Washington in a tiny community (three homesteads) called Evans, just north of the town of Colville. After a year there I moved down to Wasco County, Oregon to a town called The Dalles on the Columbia River. There, working from a small office based in a small bigfoot museum, with a staff that varied from three to five, I chased Mr. Bigfoot for another nine years. During this time I wrote what is generally regarded as the definitive work on the phenomenon, *The Search for Bigfoot,* (Simon & Schuster, NY, 1975) a book that in paperback turned out to be a bestseller.

In late 1979, new interests, mainly the prospect of running big game photographic safaris in my old hunting grounds of the White Grass Plains of southwest Nepal, again took me away from the great search and it was not until 1992 that I got back into it again.

In that year I took a wealthy mid-west gentleman and his family members on safari in the White Grass Plains. Sitting around the campfire at night, with the Himalaya gleaming in the distance, our conversation naturally turned to the yeti and then, as a natural course, to its so-called cousin, the bigfoot of the Pacific Northwest. My client was fascinated by the prospect of what he called an American yeti and I believe that he based its credibility, there and later, on two things: one, the fact that I personally believed in the phenomenon; two, the personal discoveries that I, using my expertise in sign reading (tracking) and in natural history, had made in the Pacific Northwest.

The group had an enjoyable safari with me and then went home. A little later my season came to a end—the seasonal run was October to May—and I was in the US again when my safari client called me. He asked me, just as Tom Slick had all those years before and Tom's friend, Wood, did in 1969, if I would be interested in having another crack at finding a solution to the mystery. He said that if I was, he would be willing to sponsor a search.

I thought about it, decided that I was interested, called him back and told him that I was, the result being, within a few weeks, another bigfoot project, this one to be called the Bigfoot Research Project III.

For the new project, I based my research team in the delightful Hood River Valley in northern Oregon. I bought three vehicles, a hundred thousand dollars worth of equipment, and recruited a full-time staff of five persons. I then put together an eight-person board of advisers, set up a twenty-person group of part-time field researchers, and organized a fifty-person standby team (whose job it would be to come in and assist in the event of a physical find).

I ran the project for five years until, with one million dollars spent, my sponsor decided that he had put enough money into it and I decided that I had put enough time into it, and we mutually agreed to close it down.

So, in summary of my bigfoot background and experience, by which I mean full-time professional research into the phenomenon, this is what we have: two and a half years in the sixties, ten years in the seventies and five years in the nineties; for a total of seventeen and half years in all.

Will that be satisfactory to my readers? I hope so. If it is, then let's move on from here. Let's look at how we are going to approach this problem of finding a bigfoot, of seeking a solution to this great mystery, and how we are going to go about it.

Let's start by looking at where we are going to do our searching—the area that my research suggests is the present habitat of these strange and wonderful creatures. Let's look at why and how they live there, important considerations that you need to know and which, when you do, will greatly assist you in your quest.

CHAPTER 2

Bigfoot Habitat

There are, it would appear, people who have taken an interest—a part-time interest, I should point out—in the bigfoot phenomenon who claim that the creatures are to be found in every state of the USA. For the sake of the reader who lives in, for example, Kansas or Texas or Illinois, and wants to get into bigfoot hunting in any of those states as a serious hobby, I would love to be able to support this marvelous claim. Alas, I cannot.

My reason for not being able to do so is that a simple examination of the so-called findings of these people shows them to be based on very flimsy evidence, such as second and third-hand reports of large, manlike figures supposedly seen in the wee hours on back-country roads; odd nocturnal sounds; and (more often than not) improper identification of natural signs.

I need to point out that some of these regional "findings" have been the result of serious research by sincere people within their own selected areas, usually within their own states. However, this does not make them any more genuine, or credible, and the author personally feels that claims that the creatures exist in every state of the US are quite unacceptable.

The area that my research, and the findings of my associates, suggests is the present habitat of bigfoot is the Pacific Northwest. It begins in northern California at a line drawn roughly east from the city of Eureka, and then continues north from there through two large areas of rugged terrain—the coastal ranges and the Cascade mountain ranges of Oregon and Washington—all the way to the Canadian border and then, in British Columbia (BC), on up within the coastal ranges to about as far as the southern main Alaskan border.*

The coastal range habitat, coming north out of Oregon, crosses the Oregon–Washington border and goes as far as the general area of Grays Harbor in Washington. There, for reasons unknown, it ceases and it does not begin again, within another set of coastal ranges,

---

*I realize some of the islands off the coast of BC are part of the state of Alaska. For the purpose of this discussion they are in the BC geographical region.

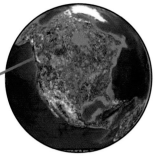

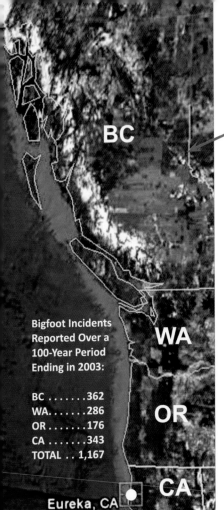

BC

Bigfoot Incidents Reported Over a 100-Year Period Ending in 2003:

BC 362
WA 286
OR 176
CA 343
TOTAL . . 1,167

WA

OR

CA

Eureka, CA

Shown here is roughly what is considered to be the true range of bigfoot. It comprises about 500,000 square miles of territory. Of course, not all of this is suitable habitat, but certainly two-thirds would be acceptable. This means that the creature has about 330,000 square miles of heavily forested territory where few people venture. Generally speaking, bigfoot has most of the region to himself. No other part of North America has the same conditions as this region—food, water, cover and space—essential for the creature's existence.

Photo: Image from Google Earth, © 2011 Europa Technologies; © 2011 Google, Data S/O, NOAA; U.S. Navy NGA, GEBCO; US Dept. of State Geographer.

until BC. Why this is we do not know. The rugged, heavily forested ranges of the Olympic Peninsula, where Washington's coastal ranges come to an end, suggest perfect habitat for bigfoot. Yet no credible evidence of the creatures has ever been found there.

Likewise no evidence, certainly no credible evidence—which is what these habitat findings are based on—has come out of the mountains of eastern Oregon or eastern Washington, areas which, including as they do the well-timbered ranges of the Blue Mountains and the Wallowas, do contain what is essentially suitable habitat for bigfoot.

Now we have the habitat and we know where to go. Next, let's decide how we are going to get there.

CHAPTER **3**

Into the Interior

We live in a wonderful country, totally modernized, with facilities that other countries envy. For example. we can fly from any city in America to any other city in a matter of hours. We can then take commuter airlines, the mighty puddle-jumpers, as they are called, to anywhere at all, including places like Stone Mountain, Georgia and Hood River, Oregon. When we get there, we can rent a car and within another hour or so be anywhere there are driveable roads and human habitat. After that?

Well, after that—and of course if you are heading for bigfoot country in the Pacific Northwest—it may be a little different because if you look at a map that shows human habitat you will see, somewhat surprisingly, that for the most part people living in the Northwest don't seem to live very much above the 3000-foot level. Is this because this is bigfoot country? Not really. The custom seems to date back to the days before modern utilities (especially electricity) enabled us to live comfortably, which was difficult to do in mountainous country where at higher elevations winter temperatures often dropped below zero. The US federal government declared much of these upper levels pretty much off limits to human occupation by claiming them as federal land and closing them to development. As a result of this, there are not many easily driven highways in the US side mountains; and if you want to go into them what you will encounter, for the most part, will be a maze of rough, narrow, gravel surfaced, one-lane forest roads, many of them with bewildering titles and numbers. The same conditions apply to Canada, although here I am not familiar with any government rulings. This is the back country of the Pacific Northwest. Back country. Off the beaten track country. Bigfoot country. Sounds confusing? Maybe a little frightening? It should not be because there is a simple remedy for access. It is this.

Before you leave the highway and enter into the great unknown of the mountain forests, or into any of the wilderness areas (regardless of how you plan to travel, on foot or by motor vehicle) drop into a United States Forest Service (USFS) or Canadian Forest Service

A partial view of some of the mountain ranges that lie 60 miles north of the city of Vancouver, BC. The image is satellite created, from a height of 19 miles. What is seen here is roughly 30,000 square miles of thickly forested, rugged wilderness. The area is roadless, meaning that all travel has to be on foot and greatly varying elevations multiply the distance that one must travel to get from A to B. Few people venture into this kind of country and those who do must be properly prepared for all kinds of hazards of rough terrain including cliffs, gorges, and fast-flowing streams that are often impossible to cross on foot. They should also be equipped to survive extreme weather conditions which, in mountainous country, can change for the worse within hours. *Photo: Image from Google Earth; Image © 2011 GeoEye; © 2011 Cnes/Spot Image; Image © 2011 Province of British Columbia; Image © 2011 Digital Globe.*

facility (there's always a Ranger Station not too far away) and pick up some good large scale maps of the area and study them. The maps cost pennies; some of them are free. Do your preliminary study while you are still at the Ranger Station so that if need be, you can ask questions of the usually very helpful people who run them. Forest Service people will always caution you as to fire hazard conditions if applicable. Pay very close attention to what they say in this regard.

Then, having decided where you want to go, make a plan with a definite itinerary and travel logistics suitable to what you see in your maps. You should follow this initial procedure—the purchase and study of suitable maps and, using them, the practical application of the logistics of travel—anywhere in California, Oregon, Washington or British Columbia.

CHAPTER **4**

Where to Start—Habitat & Passage Ways

The previous chapter discussed what is believed to be the present area of habitat of bigfoot. So, where to start? The Pacific Northwest, as well as containing great ranges of wild and rugged country, also has cities, towns and farmland; and in some areas quite dense human population centers. Obviously you want to get away from all that, because that is not where bigfoot are going to be.

So, start by studying maps that show unpopulated, forested country. Remember, cover is a prime requisite of the bigfoot and in the northwest this means forest. Then, from your maps, chose any forested area that is as far away as possible from human population centers, or that is large enough in itself to preclude their proximity. Examples are: in northern California, the Trinity Alps and the Marble Mountain Wilderness area; in Oregon, the Siuslaw National Forest and the Mount Hood National Forest; in Washington, the Goat Rocks Wilderness or the Gifford Pinchot National Forest; and in British Columbia, just about anywhere up its vast, nearly deserted coast that begins north of Vancouver and winds its way for hundreds of miles to the Alaskan border.

Having chosen a general area, move on to select a specific one. You can do this by studying the terrain, and you can start by looking at what might be passageways or potential travel routes used by bigfoot. Why passageways? Studies show that bigfoot are nomadic, and so it is generally accepted that they are constantly moving. To do this, but at the same time stay within the boundaries of their habitat, they must travel north to south, or vice versa. They are constrained in this way by what is seen on the maps as a long, north–south oriented rectangle, one that contains all of their known territory through both the US Northwest and British Columbia.

So, look again at your maps. See where physical features, like major mountains, or large lakes, as well as open country (i.e., country that is not forested and therefore does not provide adequate cover), or human population centers squeeze the northwest rectangle in certain places, narrowing it down to what might be called "necks of land."

Then think of a bigfoot traveling through this area and ask yourself what its route might be? Obviously, within that neck of land.

Now go one step further. Select one of these "necks" and look at what its terrain consists of and say to yourself, if I were a shy, elusive creature like a bigfoot, what route would I use to pass through there? What route would be safest for me and at the same time allow me passage through the area with a minimum of physical effort?

Safety for a bigfoot, of course, means being able to move without being seen and the prime requisite for this is cover. In other words, forest. So the route must have forest, and continuous forest, all the way through it, if possible.

Again, looking at the physical features of the terrain, you may find that it contains high ridges. If these are angled in the right direction, they may serve as possible passageways because this means that they will allow a traveler—you or a bigfoot—transit through the area with a minimum of effort (in this case travel with the least altitude loss or gain, an important factor in traversing mountainous country on foot). So, high ridges that cut across rugged country offer good possibilities as bigfoot passageways, and, as such, should be kept in mind when search areas are being determined.

Other passageways through rugged terrain may be provided by creek beds. The Northwest has thousands of small creeks, some of them with fairly deep water, but many of them quite shallow and with only the occasional deep pool. In many places they offer less difficult transit through an area, especially if the area has dense brush, such as alder or vine maple. The beds of the creeks themselves are always clear of growing brush and thus provide for easier movement. In my own research, I and my associates have walked many miles in creek beds, both while searching for signs and also as a means of passing through an area. The water may be cold, and one's boots will often fill with sandy water and have to be emptied. However, the alternative—hacking and pushing through dense brush, hour after hour—is certainly much less desirable.

Lastly, an area may be traversed by trails. These too offer passageways for wild creatures. However, my research suggests that bigfoot may avoid trails, possibly because of the telltale signs that walking on them might leave. Bigfoot sign—footprints—on mountain trails is very rare and has seldom been found.

CHAPTER 5

Equipment & Protective Gear

So, now you have selected your area and are ready to go in there and start your search. And if your enthusiasm is high, why, of course there'll be a bigfoot waiting for you around every corner! But whether there is or not, now you are into bigfoot territory and all ready for that first encounter. And what are you going to do when that first one pops out from behind a tree and goes Boo!? Well, we'll come to that in a minute, in the next chapter, where contact gear is discussed.

For now, most important, let's look at the equipment you have taken with you, including the gear that is going to protect you from the elements when, without warning, they suddenly become very bad-tempered and change from providing you with a nice sunny afternoon to something quite different—like a fast drop in temperature, rain and a nasty increase in wind velocity with a chill factor that can possibly be dangerous. Rapid weather changes are a common factor in mountainous country—something to keep in mind.

Essentially, regardless of whether you go by car or on foot, the basic equipment that you take with you must be designed to keep you safe. Comfort is of course desirable, but safety comes first. Thus your gear must include adequate nourishing food, water where such may not be available, quality footwear, and clothing with head protection that will keep you warm and at the same time protect you from wind and cold rain. And what about shelter? If you have a car, then your car is your shelter. If you are on foot, then you must have a tent, one that will stand up against up high wind and that is thoroughly waterproof.

That's it. That's the basics. Additional to that? There are some useful extras, and depending on how much you want to take (especially if you are on foot and weight is factor), they should include:

- two-cell flashlight with extra batteries and an extra bulb
- medical kit that includes sun block if you are susceptible to sunburn

- waterproof writing paper and a black indelible marker pen (for making records and for leaving notes in case of an emergency)
  - good clasp knife (preferably with a serrated blade)
  - general purpose multi tool (such as a Leatherman)
  - waterproof matches in a waterproof container
  - fire-starter kit
  - good compass
  - axe or a heavy bladed brush-cutter machete
  - pair of quality binoculars
  - large scale map of the area (with a scale of preferably four miles to the inch and showing contours)
  - wrist-watch altimeter (which will help you orient to map contours and thus altitudes)

and always
  - cell phone with, if possible, an extra battery, or an emergency pocket battery-booster.

That's finally it for the equipment and protective gear. Now, I personally suggest that you read as much as you can about general conditions in the area where you are going—seasonal weather conditions and average temperatures at different times of the year and at different altitudes, and evacuation procedures that may be available in case of an emergency.

CHAPTER 6

Contact Gear

Next, contact gear. And what is contact gear? It is the equipment that you are going to use to record that first face-to-face meeting with your first bigfoot.

Obviously, the recording that you make of your first contact is going to be on film or, more likely memory card. So, thank your lucky stars for the modern point-and-shoot camera, and thank them even more for the digital camcorder. If you plan to use only one, then of the two I would suggest a digital camcorder. Still pictures are good—clear and sharp—so still pictures of a bigfoot will be great to get. However, motion imagery (i.e., recorded movement, in color and possibly with sound) can't be beaten.

One of the errors that a lot of people make when going after bigfoot, and when putting together their contact gear, is to overload themselves. Try to avoid this. Fine, you can take along two still cameras, two camcorders, a tape recorder, and anything else you like. But remember, the less you have, and the faster you can use it, the better the results. Experience (that is to say contact experience) has shown us that meetings with a bigfoot can be very, very brief— usually measured in seconds. Why? We'll come to that in a minute.

So, keep your contact gear simple. I would suggest a camcorder carried where it can be used instantly and, as backup, a simple point-and-shoot still camera, also carried in a quick-access pocket. That is really all you need. Or, put it another way, all that you are going to have time to use.

Have you used a camcorder before? Used that camcorder and that camera you just bought? No? Then take time out for both of them and learn how to use them. Shoot both video and still images over and over until you can use both cameras blindfolded. Is "disaster" the word for having a bigfoot walking towards you in the woods and having to look into the camcorder manual to see what button to push? Disaster is a good word. In addition to which, of course, (a) you'll never forgive yourself, and, (b) no one is going to believe you when you tell them…"But, but, but, it was there, I saw it, believe me, it was right in front of me, believe me, please. I just could not find the right button to press. Pleeeease!"

So learn to use your contact equipment like an expert. Remember, the law of averages (of bigfoot contact) says that you may well get just one chance. One only. And that's it.

And now? Now you're all set. You're out in the woods, high up and deep in the heart of bigfoot country and all you need is that first contact. How do you go about planning for that? This comes next.

CHAPTER 7

Making Contact

The very best way to make contact with and observe wildlife, be it tiger, rhino, elephant, Pacific Northwest deer, bear, or mountain lion, is to select a place where your studies tell you the animal will hopefully come. You then sit there in concealment and wait for the animal.

Concealment? That's easy. First decide if you want to be on or off the ground, by which I mean in a tree. If you are going to be on the ground you can either use natural cover (bushes, thick grass) or, using your brush-cutter machete or axe, build something yourself. Make sure that it completely conceals you, and also make sure that you have an opening that will allow you to use your contact gear (camera and/or camcorder) with minimum movement and sound.

A tree blind is actually better, with less chance of detection by your quarry. This is because most wild creatures look along the level of their eyes and, unless movement or sound attracts their attention, seldom look above that level. In a tree, you may well be able to use the natural contours of the tree itself, especially the trunk, for a platform, thus eliminating the work of having to build such. Remember, however, that in a tree, even though there is less chance of you being detected than on the ground, you should still construct something that will give you some concealment. Obviously you are going to have to move from time to time and it is movement that catches the eye of animals. You can use natural materials for this, like leaves and grass. Otherwise use camouflaged nylon, or cotton netting (netting may be better because with movement it's less noisy).

So, now we have covered concealment at what we hope is going to be a point of contact. Where is this going to be, and why? Difficult questions? Not really, if you know the habits of wildlife.

Please note that in chapter 29, Making Contact with a Yeti, I have provided some other pointers which are equally applicable to bigfoot under similar circumstances. Please keep this in mind as you read that chapter.

CHAPTER 8

Selecting Your Place of Contact

Unlike humans, who, when they are in the forest, may have both the leisure and the desire to wander without rhyme or reason, wild creatures are different. Theirs for the most part is a life with no time for leisure, one that is designed for survival and that has to be spent looking for food, eating, sleeping, taking advantage of cover, mating and avoiding predators—including what is for many of them the most dangerous of all the predators, man.

So when a wild creature goes from point A to B in the forest, it is doing so for a reason, which will probably be one or more of the above. Leaving A with the intention of going to B, it is going to travel, as far as it is able to do so, in a straight line. Thus, if you know roughly where A is and have a reasonable idea of where B is going to be, you are off to a good start with a solution to the problem of where you are going to establish your contact point.

Next, wild creatures, just like us, will, when traveling in rough country, always take the line of least resistance. So, what happens when an animal sets out from A to go to B and on the way encounters a small lake in its path? Will it swim across it? No, of course not, any more than we would do. It will go around it. Again, assume it finds a steep, rocky slope in its line of movement, or a small cliff. Will it climb the rock slope or the cliff if there is a way around? Of course not. It will use the way around it—the path of least resistance.

Now assume that in the animal's path of direction there is a steep hill. However, set in the top of the hill there is a saddle that will allow access to whatever is on the other side, and at the same time cut down on some of the altitude that is needed to get over the hill. Will it head for that saddle and use it? Of course it will.

So, you find yourself in an area that has a cliff running across it, a small lake in the foreground, and a hill with a saddle in the top. Where do you set up your hide? It's obvious—anywhere on the natural route through the area that bypasses all three obstacles.

If this is not enough to help you decide where to establish your ambush, then look for game trails. Studying these you will quickly see what I am talking about here—how wildlife goes in a straight line as much as possible, but at the same time in combination with the path or line of least resistance.

Look for game trails. They are easy to find and their surfaces, especially when they are soft, clearly tell what wildlife is using them. Study them, see what is using them and select your place of contact accordingly.

Add to your selected place one other factor; this is what might be called the possibility of sudden human interference. If you are in a remote area this is not really likely. But keep public campgrounds in mind, or popular fishing spots on rivers, where people may be moving or camping. Why? Because all wildlife knows its own territory as well as you know your own backyard. It will know where people are liable to be, or liable to appear suddenly. It sees this as representing a hazard and will act accordingly, often to the extent of making lengthy detours around the place of potential danger. So make sure that your selected contact point is well away from anything like this—well away from people places.

Add all of this together and you should be able to make a good choice of a place from where to watch and wait for an unsuspecting bigfoot—a place of potential contact.

CHAPTER 9

What Exactly Are We Looking For?

At this point it would probably be a good thing to have a look at exactly what it is we are looking for. In other words, who or what are bigfoot and, when you make that first encounter, what exactly will you expect to see?

Descriptions by eyewitnesses tell us that essentially bigfoot look like humans. In other words they have a generally human appearance, with a torso, a head, a neck, two arms and two legs. However, there are two basic differences between us and bigfoot:

1. Bigfoot have thick and quite dense hair growth all over their bodies, while the normal human does not.*

2. Bigfoot are, in comparison with the average human, much taller and bigger.

So what we have with bigfoot, then, is a large humanlike, hair-covered creature. How large? Quite large is all that we can say for sure because whereas eyewitnesses can be depended on to describe simple appearance details like color and shape, only a professionally trained observer can estimate height and weight with any degree of accuracy. Nevertheless, based on credible eyewitness reports, a reasonable

*Note:** The human body has hair which at one time in our existence had growth and the number of hair follicles equal to that of a modern gorilla and also, it would appear, to the hair growth seen in a bigfoot as described by eyewitnesses. However, we started wearing clothing thousands of years ago, which same, in addition to the use of fire for warmth, eliminated our need for hirsute body protection.

estimate of height for an average adult male bigfoot would be between six and a half and seven feet, and for a female, four to six inches shorter. A reasonable estimate of weight would be (for the same average male) something in the region of four hundred and fifty pounds, and for a female, about fifty pounds less. (Note: For more discussion on size and weight, see chapter 14, The Evidence: The Patterson/Gimlin Film.)

Size, then, essentially separates us from bigfoot. They are on the whole much bigger than we are. Of course we do have human individuals, like some of our "giant" basketball players who, at seven feet plus, are equal in height to a bigfoot. And occasionally we have a single individual like the huge Frenchman, Andre The Giant, as he was known, who (if we are to believe his agents) stood seven feet four inches and weighed 460 pounds.

One other thing, again described by eyewitnesses, makes bigfoot a little different from us—the shortness and thickness of the neck and the setting of the head on the neck. The bigfoot neck is described as being very short and thick. So short, in fact, that the position of the head, in relation to the body, seems to be half way below the line of the shoulders. We humans, because of the length of or necks, are able to turn our heads close to 180 degrees and thus, using our peripheral vision and without turning our bodies, look behind us. A bigfoot may not be able to do this because of the shortness of its neck. In fact, several eyewitnesses have reported seeing a bigfoot turning the whole of its upper body to look back at them when, given it had the neck mobility of a human, would probably only have turned its head.

CHAPTER 10

Are Bigfoot Human?

So if size and being hair-covered and having a short, thick neck separates us, at least by description, from bigfoot, then what are the characteristics that move it closer to us and to being human? We have already discussed its humanlike shape, although this is only a marginal consideration because most primates (human and non-human) have the same general shape. Nevertheless, that it appears to be a primate, puts it in the same classification as humans. Beyond shape, there are two other major aspects: the way it walks and its facial features. Let's take a look at these aspects and see how they suggest that the creatures are hominids (rather than hominoids), possibly even a very early form of human. We will begin by thoroughly examining the way bigfoot walk. It is in this, their form of what scientists call locomotion, that we see a very definite clue to the creature's place in the chain of evolution.

Our world contains all kinds of wildlife. We have mammals, birds, reptiles, fishes and insects. All of them move around much of the time and to do so they use their own particular form of locomotion. Generally speaking, mammals walk, birds fly, reptiles wriggle and slither, fishes swim, and insects hop, jump and fly. Bigfoot, of course, is a mammal, and so are humans. In this discussion I want to exclude the latter, so what we have are non-human mammals.

In North America, all non-human mammals are quadrupeds.* That is to say, they walk on all fours. Bears have been seen walking upright, on their hind legs, but only for very short distances because this is not a natural form of locomotion for them. The same applies to the wild primates of our planet, apes and monkeys, which are all, without exception, quadrupeds. Indeed, upright stance and bipedal locomotion is as foreign to them as walking on all fours would be to us.

So, is there any living organism in North America that walks

* I specified North America for this statement because this is the region I believe bigfoot exist. However, it is essentially correct for all non-human mammals with the exception of the kangaroo and wallaby.

upright, on two legs, any creature for which this is a natural form of gait? Answer, yes. We humans. And for us this is our natural form of locomotion, one which we as hominids have used for millions of years.

In studies that I and my worthy associates have performed over many years, every single report of visual contact with a bigfoot described the creature as walking upright on two legs. There are credible reports of bigfoot sitting and crouching, but no one had ever seen one walking on all fours. This fact strongly supports the hypothesis that bipedal locomotion—walking upright on two legs—is bigfoot's natural form of gait.

We will now consider facial features. How do they support the idea that bigfoot is human? There are some very good reports with descriptions of the creature's face as seen by reliable witnesses, not at night by car light, not on some distant ridge with the sun going down, but in good light and within short distances.

Many of these reports talked of a "strangely human face," or a "strangely manlike face," with a straight nose, large eyes, white areas (called the sclera) around the eyeballs—a very significant factor, discussed below—and with the face bare of hair except in those places where hair is common in humans (i.e., on the sides of the face, on the lower jaw and on the top of the head). None of the credible reports that we studied mentioned a gorillalike or apelike face.

As to the white sclera. What does this mean and why is it significant. Simple. Among the scientifically acknowledged living animals of this planet on which we live, only humans have a continually visible white sclera (whites of the eyes). Some animals have a white sclera, but it is only seen when the eyeball (pupil and iris) is moved to the extreme right or left sides, or is in the extreme up or down position. Ever look at a dog's eyes? A cow's? Check them out and you will see exactly what I am referring to.

What does this mean? It means that bigfoot, in addition to sharing bipedal locomotion with us, also shares having a continually visible white sclera—two important features, one unique to humans.

So what is the creature you are hopefully going to encounter in that first meeting? My guess is that what we are looking at in these wonderful primates is a very primitive form of man, extinct throughout all of the world with the exception of the Pacific Northwest. A hominid form that has, like us, come down through the millennia with very few physical changes. Like us they walk on two legs and they have continually visible white sclera. But unlike us, and for reasons not quite understood at this time, they prefer to stay wild, to live without clothing or shelter or tools in one of the most rugged areas of the planet.

CHAPTER 11

Bigfoot Sustenance

Among people who have made studies of bigfoot, it is fairly well accepted that in the matter of food they are probably omnivores. The word omnivore is from the Latin "omni" meaning everything or all, and "vore" meaning eating. An omnivore therefore is a creature that, like us, eats everything.

Some wild creatures, like deer, are herbivores, living solely on vegetable matter. Others, like the cats, are carnivores, eating only meat. Some, like us, and like bears, as an example, are omnivorous and this is very probably what bigfoot are.

If they truly are omnivores, then they will have no trouble finding food in the Pacific Northwest, the vast forests of which contain hundreds of edible plants, including wild berries, wild onions, wild asparagus, edible ferns, seeds of all kinds and flowers, such as wild roses. The lakes and wetlands, of which there are many, have edible sedges, bull rushes and water lilies. The mountain streams abound in fish, frogs, and aquatic insects. And as carnivores and some omnivores usually will also eat carrion (dead animals), these same streams will annually provide a veritable harvest of fish—not the least of which are the bodies of salmon after they have spawned and died.

Among the eyewitness reports that relate to food, we have one from a Canadian engineer who encountered a bigfoot carrying a fish in its hand. We have another from two Canadian surveyors who saw two of the creatures standing waist deep in a small lake pulling up edible water lily stems and eating them. Another man watched one eating huckleberries, picking them off a bush one by one and popping them into its mouth.

What is the purpose of these examples? If nothing else, it's to refute the argument of skeptics who mistakenly claim that for something the size of bigfoot there simply isn't enough food in the forests of the Northwest. I say, tell that to bears—big, ever-hungry animals that just about never stop eating. They seem to survive very well and there are thousands of them in the Northwest forests.

CHAPTER **12**

Bigfoot's Skills

In their natural forest habitat, bigfoot seem to exhibit unusual wood-craft skills. They obviously know how to use cover to their advantage; they probably move very quietly and they are, pretty much like all wild creatures, very wary and alert to the presence of man. Movement by a bigfoot through an area will probably include staying on high ground such as ridges—using the long range visibility factor that elevation gives—or via deep ravines to take advantage of the cover they provide. Either way, my feeling is that their field craft will be first class—much of which will be designed around staying out of the sight of man. For whereas many animals of the Northwest forests are subject to predation, the only predator that is of any danger to bigfoot is us. It is possible that they sense this, for they avoid man as any shy forest creature does; the paucity of sightings supports this.

Mind you, the average man is not too difficult to detect when he is in the forest. In fact, for the most part he sticks out like a sore thumb. I have been out in the Northwest forests with people who smoked, coughed, talked in loud voices, and at every advantage seemed to make as much noise as they possibly could, including slamming car doors and crashing through the brush like wild buffaloes. There have been times, in the company of people like this, when I have imagined myself as a bigfoot, standing on a ridge half a mile away from the epicenter of the disturbance and saying to myself, yes, time to move on, time to fade away.

Again, the creatures, although we are not certain, may also have a fairly keen sense of smell. Many forest-dwelling creatures, as opposed to those that live in open country, have well-developed olfactory capability and the bigfoot may be among these. It is useful to remember this and to keep in mind that certain strong smells, such as that of tobacco smoke, can be detected by wild creatures up to distances of 400 yards from its source. Cigarette and cigar smoke will also impregnate leaves and grass in the immediate area of its emission with an odor that will remain detectable for days.

I personally have a reasonably well-developed sense of smell (I have never smoked, which helps). Walking up a creek bed in northern California some years ago, I detected the smell of pipe tobacco. Within a few yards I came on a small campsite where a man had spent several days camping and, it was obvious, enjoying his pipe. Later I learned that when I smelled the tobacco smoke, the man had been gone from there for seventy-two hours.

As to hearing, like most forest creatures, that of bigfoot is probably very well developed. Visibility is limited in forest, especially in dense forest, and to compensate wildlife for the visual loss imposed upon it by dense vegetation, nature automatically endows it with additional ability in one or more of its other senses—in this case hearing, as well as a keen sense of smell.

Two examples of what might be called compensatory sensory abilities, from the author's safari days experience, are to be found in two separate species of deer and antelope. One is the Asian swamp deer, an animal that spends much of its time in very dense, high grass, where the visibility is often no more than ten feet. To compensate for its inability to use its eyesight to detect danger, nature has given this animal huge ears, sensory appendages that provide it with acute hearing capability. On the other hand the Asian black buck, a plains antelope that lives in open country, with excellent visibility, has a poorly developed sense of smell. To compensate for this, nature has provided it with excellent long-range eyesight.

Probably good eyesight, a good sense of smell, excellent hearing and superb alertness make bigfoot a fairly formidable adversary in the forest. In other words, if you go blundering along making a lot of noise, you are certainly not going to bump into one. Quietness is the watchword in the forest, combined with as much stealth as you can manage. Walk softly, stop and listen quite often, stay in cover as much as possible, and if you have to communicate with a companion, whisper or use hand signals. Also remember, as I have outlined here already, that the best way to make contact with a bigfoot is the same way that is most successful with all shy wildlife—from concealment in a blind, or hide, sitting quietly, with minimal movement and, if possible, off the ground in an elevated position.

CHAPTER 13

Claimed Paranormal Aspects

Some bigfoot hunters credit the creature with extrasensory perception and interdimensional capabilities. In the field of extrasensory perception, claims have been made that bigfoot, without using any of their normal senses, may be capable of being aware of the approach of a person long before that person is seen. Also, to know the person's thoughts and, if necessary, even communicate with the person using telepathy. Additionally, this expertise, it is said, enables bigfoot to influence a person's emotions and thinking to the point of being able to persuade the person to turn around and retrace his or her steps, long before the person might actually come into viewing distance of the bigfoot.

I do not contribute to this thinking. My experience with wildlife suggests that no animal is capable of this and I feel that, in the case of those who expound it as a theory, it is nothing more than pure speculation and distorted imagination.

As to the interdimensional theories that claim bigfoot are capable of appearing and disappearing at will, with an ease that is beyond human understanding, again I do not believe in such. I do not feel any wild creature, anywhere, has this capability and certainly not bigfoot. How this thinking has come about, I believe, is the result of the inexperience of the claimants in the natural ability of wildlife to take advantage of cover and use it. To be sure, at times wild creatures do *seem* to appear out of nowhere and then, within seconds, to the astonishment of the observer, disappear again. I believe that wild creatures (certainly those which are subject to predation) are capable of extraordinary abilities when it comes to avoiding danger. However, this is based on nothing more than a refined and well-developed form of alertness and awareness built upon their survival instincts. I personally feel, after nearly fifty years of research into the bigfoot mystery, that if these unfounded claims were true, then I would have encountered them myself at one time or another. I never have.

As to the extraordinary natural ability of wildlife to use stealth and camouflage to evade detection, I will provide an example. Once, in a forest in Washington State, I was sitting on a small bluff, looking down into a little valley. Below me was a patch of thick brush, mostly alder and vine maple. I had a clear view into it and had been sitting and watching it for an hour. Suddenly I noticed a small movement and, as I watched, fifteen elk that had been resting in the scrub got up right in front of me, shook themselves and then walked away. I (with excellent eyesight and much experience in its use with wildlife) until then had not seen as much as a hide of one of them.

In various areas of the Pacific Northwest I have seen bear and mountain lion seemingly appear and disappear in the same way. In Asian forests I have even seen a 16,000-pound elephant do the same thing—literally vanishing from sight within seconds. While these disappearing tricks may seem almost miraculous to some people, persuading them that there is interdimensional capability involved, it is nothing of the sort. It is the natural ability of wildlife to take protective action when danger, or the possibility of danger, threatens. Bigfoot undoubtedly have this ability as well.

So, don't be alarmed if a bigfoot suddenly appears right in front of you. He did not just step out of another dimension and he is not about to beam you up to some distant planet. He is as ordinary a creature as you are, and all he is doing is practicing his woodcraft, something that is as natural to him as walking is to you, something that protects him from the threat of danger (mainly, of course, us) and at the same time contributes to the survival of his species.

CHAPTER 14

The Evidence

If you have got this far you have probably noticed that I talk about bigfoot as though I am 100 percent certain that they exist. I talk about people seeing them, about their size, what they look like, their food habits, their woodcraft and more. You might question this, as you should, and say to yourself, what gives this guy the right to do this? What gives him the authority to say, "Yup, they're out there, I'm pretty sure of it." Where's the foundation for all of these statements; the evidence? Where's the credibility and the background? Is there a background and if so, isn't it time in a guidebook that it was put on the table for inspection? I think it is time, so here it is.

The evidence that supports the credibility of the existence of bigfoot comes in five parts: the documented history, North American Native legends, footprints, sightings (i.e., eyewitness reports) and a little piece of 16mm film taken in 1967 that is known to bigfoot researchers as the Patterson/Gimlin film.

Documented History
As evidence, history is particularly significant because of one thing—it is documented and it is factual. This documentation, much of which was unearthed by myself and the many fine people who worked with me during my three bigfoot projects, is mostly found in old newspapers and magazines, and in the letters and diaries of the early settlers—including miners and missionaries—of the Pacific Northwest.

What is significant about the documentation is the fact that always, without variation, it describe what could only be a bigfoot. No variation? Well, there is some and this is, quite naturally, in the areas of color and size. As has been stated, bigfoot do seem to vary in color somewhat, from dark brown to black; and obviously they vary in size. But apart from that, all the way back to the very earliest report that we could find (from the *London Times* in 1774), the unvarying description is of a large, upright-walking, hair-covered, manlike figure. If this is not the bigfoot of the Pacific Northwest, then what else could it be?

A history, as a background to anything, is very important. In the case of the bigfoot phenomenon it serves to show that it is not something that evolved in recent years with the discovery of footprints in the Northwest. It goes to prove, certainly as far as this researcher is concerned, that the phenomenon has been with us, and has been documented for at least 200 years.

North American Native Legends

Supporting the documented background are the Native American legends, and it is fascinating to note that every Native tribe of the Pacific Northwest has a name for bigfoot in its own language. Not only that, but the names all mean pretty much the same thing, which essentially is, Big Man, Big Man of the Forest, Giant Man of the Forest, or Hairy Man. To the Salish people bigfoot are the sasquatch. To the Hoopa and Yurok, the omah, and in my files I have thirteen other Native names.*

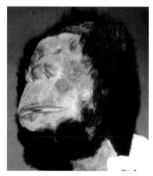

This mask, carved by a Native from Chehalis, BC, is believed to have been inspired by an actual bigfoot sighting.
Photo: C. Murphy

I have interviewed many Native people in my research and while few of them claimed to have actually seen a bigfoot, all of them stated quite positively that the creatures exist. They said they believed that in the past there were many more of them, but that in recent years they seem to have diminished. Why, they did not know.

Footprints

The footprint of a bigfoot looks very much like a large, unusually wide, human foot imprint. The creatures have five toes on their feet and their footprints usually, but not always, show all five. Sometimes only four will show and this is because the little toe does not always connect with the ground with sufficient force or weight to make a noticeable impression, just as in a human print.

*Recent research by others has revealed a total of 124 different Native names that are, or may be, associated with bigfoot.

Bigfoot footprints, obviously, vary in size—ten inches to fifteen inches is about the norm. Over the years I have personally seen five sets of footprints, two of which I found myself. These varied from twelve inches to fourteen and a half inches.

As to footprint width, in comparison with a human print, the bigfoot print is unusually much wider. This is probably to compensate for the considerable weight of the creature that made it.

So, can a big human footprint be mistaken for a bigfoot print? Yes, except for one thing. In a human footprint one sees, just behind the big toe, the impression of what looks like a small round pad. This is made by the muscle which is part of the propelling mechanism of the foot. In the bigfoot there are two of these, one behind the other. Have a look at my photograph of a bigfoot print (page 62) and you'll see what I mean. Keep this in mind when you find and examine humanlike prints of any kind—the second bulge is what to look for. This is what distinguishes a bigfoot footprint from a print made by the creature's nearest relative, a human.

Sightings

Now we come to sightings, that is, eyewitness reports, and with the exception of the Patterson/Gimlin film, we've pretty much kept the best for last.

An eyewitness report is a written or verbal account of visual contact with a bigfoot. Over the years in which I have taken a part-time interest in the bigfoot phenomenon, and during my professional searches, I have interviewed hundreds of people who have claimed to have seen one or more of the creatures. Believing that research in this vital area of evidence should be conducted very professionally and with great caution, my results show what is regarded by some as an overly conservative figure of credible reports. However, out of the large number of original reports, the vast majority were dismissed because the information was inconclusive or the witness was considered unreliable. In the end I finished up with just 125 reports from people whose accounts were, in my opinion, completely credible.

The credibility of an eyewitness report is made up of three contributing factors: (1) the general reliability and integrity of the person making the report; (2) the circumstances of the report including where it took place (meaning a suitable area and not, say, within a town or city limits) and at what distance (meaning within a visually appropriate distance); and (3) the duration of the sighting.

All of the 125 people who claimed to be bigfoot encounter eye-witnesses and who were subsequently interviewed by me, were mature, rational and thoroughly believable people. Among them were:

- three Pacific Northwest deputy sheriffs
- a Portland, Oregon attorney
- a Hood River, Oregon judge
- a schoolteacher
- two U.S. Forest Service forest rangers
- a pair of Weyerhaeuser engineers
- a Canadian government meat inspector
- a pair of Canadian surveyors
- an eighty-five-year-old Oregon farmer and hunter
- an Oregon State policeman.

All witnesses described encountering a large, manlike, upright-walking, hair-covered primate. Some of them remarked on the face as being strangely human. Some of them noticed how short and thick the neck was. Some of them described the coloration of what they saw as dark brown or black. In one case, the eyewitnesses were members of a four-person group, with each corroborating the other's story.

As to the circumstances of each report, the facts were totally supportive. In other words, the encounters were in remote areas—terrain generally regarded as potential bigfoot habitat—as well as which they were made at reasonable distances that allowed for careful surveillance and identification. As to the duration of each sighting, this too fitted into the credibility factor; in other words, they were of a sufficient period of time to allow for deliberate and calculated observation of the object in question (in some cases the viewing lasted for minutes).

Who or what did all of these people encounter? For that you'll have to wait to the last chapter because right now we are going to discuss what bigfoot researchers call the Patterson/Gimlin film.

The Patterson/Gimlin Film

In October 1967 two men from Washington state, Roger Patterson and Bob Gimlin, set out from a camp in the Six Rivers National

Forest in northern California to look for a bigfoot. They rode horses, behind one of which they trailed a small packhorse on a lead. From their camp they chose a route that took them up the bed of a shallow creek. They had been riding in the creek bed for about an hour when, rounding a corner, they came across what they described as a female bigfoot. The creature, they said, was squatting on its haunches at the edge of the creek and had a big, wet stone in its hands, the bottom of which it was licking.

Roger Patterson from a painting by Chris Murphy. *Photo: C. Murphy*

At the sight of the creature the pack horse reared, broke its rope and, turning, went galloping off down the creek bed. One of the riding horses did the same thing, throwing and rearing down on its rider, Roger Patterson, temporarily trapping his leg. The second horse remained in the control of its rider, Bob Gimlin.

Bob Gimlin in 2003. *Photo: C. Murphy*

When the bigfoot saw the horsemen, it stood up and took a few uncertain steps away from the edge of the creek. It stopped and stood for perhaps a couple of seconds and then started to walk purposely up a big white sandbar that bordered the creek.

In the meantime, Patterson (the cameraman of the two-man team), who was temporarily trapped, managed to get his leg out from under his horse, and also managed to get his 16mm Kodak K100 movie camera out of his saddle bag. Letting his horse go, he ran towards the creature filming it at the same time. The piece of

The creature in the film is seen here when it turns and looks at the cameraman. There are other small details that might be apparent, however, the distance of the creature from the camera (about 102 feet) does not allow positive identification. *(Photo: R. Patterson/Public Domain.)*

film that he shot begins with the creature walking away from the creek and turning right along the forest fringe. It then captures the moment when the creature turns and looks at the cameraman. From that point we see the creature more or less from behind as it eventually disappears into the distant forest.

The film footage shows a large, upright-walking, hair-covered primate. Large breasts indicate that it is a female. She walks with a smooth, swinging stride and when she turns to look at the cameraman, she turns her entire upper body. When she does this, her face is seen in semi-full view. It is a hominid face, with a straight nose, and deep-set eyes of which the white (sclera) of one eye is visible. The face and body appears to be covered in short hair (not fur) which is somewhat patchy in spots.

Its arms are long, but not unusually so, and it appears to have large hands. Enlargements of film frames reveal that teeth may be faintly visible between its lips. Its feet are large, but in proportion to its body size.

Its head—and here we see something possibly very primitive—has what appears to be a sagittal crest (high bony ridge). This same feature is seen in some of our modern primates and may have been part of the cranial structure of prehistoric man. However, in the case of the subject of the film footage we are not certain as to what exactly we are seeing. It could be bone or simply hair.

I was not in the US when this footage was shot and it was some time before I was able to get to the actual film site. However, when I did get there, the physical site was essentially the same as shown in the film frames. I took a young man with me to act as a model for the creature in photographs. Using enlargements of stills taken from the film footage, I made my own calculations of the height and weight of the film subject. I performed a series of calculations that eventually gave me an accurate measurement of the creature's height, as well as something close to a reasonable estimate of its weight.

The creature seen in this extraordinary footage is massively built, stands seven feet, three and a half inches in height and has a waist measurement of seventy-two inches. These figures, which I believe are accurate, are based on simple mathematical formulas, using my model, a person of known height and weight, for proportional comparison.

The weight estimate of the film subject is arrived at in a different way and is a calculation based on my own experience in the observation of large, wild animals in both Asia and East Africa. It is, give or take a few pounds, between 400 and 450 pounds.

The creature's footprints were impressed up to about one inch into the ground. They were filmed by Patterson, and plaster casts

This photograph shows my model with a measuring pole standing at the precise spot where the creature turned and looked at the cameraman. I have my camera at the same distance as that when the film was taken. Various trees, logs, stumps, and other debris can be matched exactly with the film frames. *(Photo: P. Byrne.)*

were made of two of the prints (right and left foot). The casts measured about fourteen and a half inches in length. Three days later, some of the prints were photographed by a timber management assistant. Nine days after the filming, a researcher took casts of ten of the prints. Subsequent research indicated that the creature's pace, heel to heel, was on average forty-two inches.

Of great interest was the discovery that one of the photographs I took of a fourteen-and-a-half-inch bigfoot footprint in the same area in 1960 (seven years before the Patterson/Gimlin film was shot) proved to be identical to the film creature's footprints. This was verified by an anthropologist, Dr. Grover Krantz, at the Department of Anthropology, Washington State University, Pullman, Washington. In other words the huge female that is the subject of this historic piece of film was living in the area while I was conducting my own research there in the early 1960s. (See page 62 for a photograph of the print I discovered.)

CHAPTER 15

Summing Up

And so, you have the evidence—history, North American Native lore, footprints, sightings and the Patterson/Gimlin film.

The history is very important. Without it I would be bothered. Phenomena does not just spring up out of the ground and start off in a single day. It has to have a background; it has to have a foundation. The bigfoot phenomenon is firmly supported by its history.

The same applies to Native lore. It would be very odd indeed if our Native people knew nothing about a large, hair-covered primate that shared the forests with them. But they knew of it alright. They still know of it and they have named it, not as part of their folklore, but as a real living entity of the Northwest mountains.

The footprints? Well, I have to admit that footprints can be faked and because of this their place in the chain of supportive evidence may be slightly flawed. However, I personally have tracking skills that I developed as a professional hunter many years ago, and although I have seen some faked prints (made by the jokers who get their fun in trying to fool others) I have also seen five sets of massive footprints, all discovered in remote areas far from the beaten track; all deeply imprinted into the soil by pressures indicating great weight. Only a very big, bipedal primate could have made those prints; I will bet my reputation on that.

In the matter of the Patterson/Gimlin film footage, I think that it is important to say that both I and other dedicated bigfoot researchers have spent many, many hours looking for flaws in the film and looking, as they say, for the "man in the fur suit." Space does not allow me to include details of the intensive research that I have done on the film footage, most of which is contained in a report that is actually larger than this little book. Suffice to say, to date no one, anywhere, has been able to conclusively prove the footage to be faked. My opinion? I think it is real and, as such, of historical value.

As to the eye-witness reports, I found the 125 to which I personally gave a green flag very convincing. And their credibility, to me, was much enhanced by the character and integrity of the people who made them. Men and women of impeccable character, sensible, intelligent, clear-thinking, rational people not given to fantasizing, and in all cases people who had absolutely no motive of any kind to manufacture fraudulent stories.

And so, if what these eyewitnesses saw was not a bigfoot, then what was it? Men in fur suits? Very big men who have been doing this—making these one-man, on-stage appearances—for 200 years? Men who, let's face it, would have to be crazy to do this in any case, especially in Pacific Northwest back country where a very large number of people carry guns and where getting shot, if only by accident, would be a distinct possibility.

But if it's not men in a fur suits, what else could it be? Wild imagination on the part of all these very sane, very sensible people? A huge conspiracy of jokesters? A great conspiracy of lies among a group of people 90 percent of whom did not know each other? Or is it a large, hair-covered, very primitive human, just waiting for scientific discovery? You be the judge.

CHAPTER 16

Go Get 'Em!

You want to find a bigfoot. You want to make contact with one, get still pictures, shoot video. Well, now you have the formula, the plan, the whereabouts and the know-how. So what are you waiting for? Want to go back over a few things? Look again at the essentials? What to do and what not to do? Okay, let's do that and let's start with where to go and then go on from there.

First off, do you recall the proper approach to going into the great forests of the northwest? Find out what the regulations are about entry, about camping and about fire. Check in with the local Ranger Station and talk with the people there. They are always helpful for they are as concerned that you enjoy your hiking or camping trip, or your bigfoot expedition, as you are. They are also concerned for your safety. In this connection, check and double-check your safety gear. The Northwest forests are beautiful and a well-organized visit will provide you with an unforgettable experience. However, they can also be very unforgiving and they will prove this to you very quickly if you have not made proper preparations, and if you have not equipped yourself with essential safety gear including, always, a cell phone.

Where to go? Once again, start in the Pacific Northwest where I told you to, north of that east–west line drawn through northern California, up the rugged ranges of the Cascades of Oregon and Washington or up the beautiful coast ranges. From there, go on into British Columbia, all the way to Bella Coola. Remember what I said about the area north of Grays Harbor in Washington, up through the Olympic Peninsula to the Canadian border and how no credible evidence of the creatures had been found there? Keep that in mind and also keep in mind what we looked at in southeastern Washington and northeastern Oregon—no credible evidence in those areas and thus, as this suggests, a very limited possibility of habitat.

What else for bigfoot habitat? Look for those geographical "waistbands," those areas where the great broad chain of the mountains narrows. And keep in mind the possibility of passageways physically formed by rivers, cliffs, high mountains and lakes. Remember, bigfoot are believed to be nomadic, which means that they travel a lot. If they do, because of the layout of the land that is their habitat, they must travel north to south, or vice versa, because traveling west to east would put them in regions where there is open country that has no cover (very quickly in both Oregon and Washington); and traveling east to west would mean that they would very soon find themselves swimming in the ocean towards Australia!

And speaking of the ocean, don't neglect the beaches. For an omnivore, which is what we believe bigfoot to be, beaches are a vast source of food in the form of shallow-water fishes, crabs, clams, mollusks and other easy-to-find, edible marine life. Furthermore, if bigfoot just might happen to be concerned about those big footprints they must leave behind them when they walk, they conceivably might be clever enough to know that every time the tide comes in, which it does twice a day, it pretty much erases all the signs of anything that has been on the beach within the previous six hours.

Having decided on your general area, now take a look at a specific one.

Remember what we said about ridges, and how wild creatures will use them to facilitate travel, and also, by the way, for the security advantages provided by being able to stay high and thus keep an eye on what is below them.

Remember what we said about game trails, and how wildlife will use them to move through an area. And also about the alternative to ridges, which are deep ravines with dense cover. Study your maps, see what the terrain has to offer and pick your place accordingly.

Do you remember about the advantages of elevated concealment and how most wild creatures seldom, if ever, look up? And thus the importance of building your blind, or hide, up off the ground? Don't forget this. You don't have to be very high; fifteen feet up is fine. Just remember to keep as quiet and as still as possible. Movement, as well as sound, is what attracts the immediate attention of wild creatures.

And lastly, what did we say about the bigfoot being gentle and peaceful creatures? They are, make no mistake. You may hear stories about people being attacked by bigfoot, about children being chased and women being kidnapped. If you do, take the time to challenge the person who spins these yarns; ask for the proof. You will quickly find that there is none. For the most part, people who invent and then spread these ridiculous stories do so to excuse their brutal theory that the best way to prove that bigfoot exist is to shoot one. To me, shooting a bigfoot would be like shooting something as peaceful and harmless as a mountain gorilla, or a chimpanzee. As a young boy said to me not so long ago after a school talk, suppose some crazy like that does shoot one and the one he shoots is the last one!

Good luck, good hunting, and if you follow the directions this professionally written guide gives you, with a bit of luck we'll be seeing you on the morning news one of these days…soon after you make that first contact!

As a matter of course, I need to mention that there are many people involved in the search for bigfoot and various get-togethers and conferences are held. There are numerous bigfoot websites that will enable you to keep up to date on findings and upcoming events. The following photograph is from the Beachfoot camp-out held in Oregon in 2009.

Author (in khaki safari jacket) meeting new and old friends around a roaring camp-fire at an interesting and enjoyable gathering of bigfoot enthusiasts known as Beachfoot and held annually on the Oregon coast. *(Photo: C. Murphy.)*

Bigfoot Photo Album

In this section I present various photographs related to my bigfoot research. I have started with the brochure I prepared and distributed while operating the Bigfoot Research Project. It is essentially a summary of the bigfoot issue, aimed at encouraging people to come forward with their bigfoot experiences. Numerous cases were reported and investigated, providing me with an opportunity to meet many people whom I believe were both honest and sincere in relating their bigfoot encounters.

The Bigfoot Research Project

The Bigfoot Research Project

The Bigfoot Research Project is a benign, scientific investigation designed to prove the existence of a large, bipedal, hair-covered hominid believed to be living in the forested mountain ranges of the Pacific Northwest. The project, conducted in association with the **Academy of Applied Science**, Boston, MA, is the most professional and sophisticated approach ever attempted on the subject and the only one of its kind in the country. Over thirty years of experience provides the background essential to the professional research and methodology of this unique program.

Techniques and procedures used in the Bigfoot Research Project include the compiling of both current reports and historical evidence; the statistical computer analyses of these will provide the foundation essential to intensive field studies. Through this latter is envisioned the primary objective of the program which is, with appropriate documentation, contact and communication with one or more of these extraordinary hominids.

The Evidence

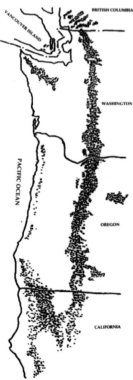

The evidence that supports the reality of the Bigfoot has its foundation in four basic areas:

> **History.**
>> **Sightings.**
>>> **The 1967 film.**
>>>> **Footprints and other finds.**

History. The history, in literature, is found in old newspaper reports, the journals and letters of the early settlers and prospectors of the Pacific Northwest, and archival records such as those of the Northwest Fur and Trading Company. The earliest known report is from the London Times and is dated 1780. Among the native people of the Northwest, of course, the record is much older, and indeed, every Indian tribe has a name for the hominids in its own language - Sasquatch, Omah, Tsonaqua, Bukwas, Selatiks, to name but a few.

Sightings. Credible sightings are a vital part of the evidence. Sightings are rare, only a few each year, indicating extreme wariness on the part of the creatures combined with a skilled ability to use the physical features of their natural habitat for concealment. Again, there may be fewer of the hominids than we think.

The 1967 Film. *This short piece of 16mm film, shot in northern California in October, 1967, is to date the only believable photographic record of a real, living Bigfoot. A recent discovery by the Bigfoot Research Project, of the similarity of a footprint found in the same area six years prior to the shooting of the film sequence, further supports the credibility of this unique footage.*

Footprints and other evidence. Footprints, averaging 14 to 15 inches and made by a large, bipedal hominid at least six feet in height and weighing up to 500 lbs., continue to be found. Recently some were discovered that contained dermal skin patterns (like fingerprints), a significant factor in hominids. Other evidence, including strong odors and powerful screaming roars, further support the reality of a group of large, unclassified primates, living undiscovered in the mountain forests of the Pacific Northwest.

What is known to date

Although many years of research into the Bigfoot phenomenon now lie behind us, much of what we believe we know comes mainly from prudent speculation. The reason for this lies in the extreme difficulty in actually studying the creatures in the vast and formidable terrain that is their habitat. (Since W.W. II seventy-three aircraft have crashed in this area and remain undiscovered.) However, these deliberations, combined with intensive studies of sighting reports and footprint finds lead us, at the least, to the following conclusions:

General:	Nomadic, nocturnal (with daytime activity only as a result of disturbance)
Habitat:	Forested mountains of the Northwest
Eating Habits:	Omnivorous
Height:	Over 6'0" average
Weight:	Up to 500 lbs.
Strength:	Far in excess of man, very powerful
Eyesight:	Superb visual proficiency
Smell:	Highly developed capability
Hearing:	Exceptional acuity
Demeanor:	Inoffensive, shy, non-aggressive
Locomotion:	Bipedal

Indications are that the creatures are more man-like than ape-like and that they possess unusual intelligence, on a level possibly close to our own. An example of this is something that our studies have discovered, that they are aware that their footprints may attract unwanted attention from man, the only entity that represents a danger to them–and as a result they hide them. This is something not found in the animal kingdom. This extraordinary hominid may well be the mysterious wild man that the great Swedish scientist, Carolus Linnaeus, (1707-1778) called Homo Nocturnus, the Man of the Night.

We need your help!

Help us solve one of the most *fascinating and popular mysteries of our time by joining* the growing number of people who contribute credible information to The Bigfoot Research Project. All information is treated confidentially and is used solely for the purpose of benign research. Sightings, footprint finds, reports of sounds relative to the phenomenon, all contribute to our computerized search for behavioral patterns. The Bigfoot Research Project considers all information valuable and invites you to become a member of the team determined to solve this great mystery.

-To strive, to seek, to find and not to yield. (Tennyson)

The Bigfoot Research Project

Brochure used with permission of the Academy of Applied Science.

The Bigfoot Information Center and Bigfoot Exhibit, situated in The Dalles, Oregon. It was a scientific, educational and research facility aimed at resolving the bigfoot issue. *Photo: P. Byrne.*

The author, driving his bigfoot research project Jeep with his friend Tom Slick. Sitting between them is Tom's younger son, Charles, and in the back, wearing a navy blue sweatshirt, his older son, Tom Slick Jr. The photograph was taken in the Bluff Creek watershed, in the Six Rivers National Forest, northern California, in 1962. *Photo: P. Byrne.*

Tom Slick, sponsor of the 1960–1962 Bigfoot Research Project, and his son, Tom Jr., bigfoot hunting in 1960 with the author in northern California. *Photo: P. Byrne.*

Author (right) with Steve Matthes, a professional mountain lion hunter with U.S. Fish & Wildlife. He took a year's sabbatical to work with the author during the 1960s Bigfoot Research Project. *Photo: P. Byrne.*

Fifteen bigfoot footprints found by the author's northern California Bigfoot Project team in 1961 on the Onion Mountain road in the Six Rivers National Forest. The prints were about 15 inches long. *Photo: P. Byrne.*

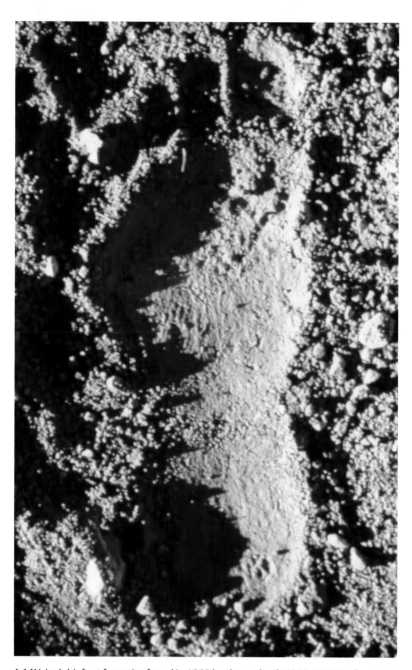

A 14½-inch bigfoot footprint found in 1960 by the author's 1960–1962 Bigfoot Project team on the Onion Mountain road (Bluff Creek watershed) in the Six Rivers National Forest, northern California. It was one of approximately 300 prints.
Photo: P. Byrne.

A 14½-inch bigfoot footprint found by the author in 1961 in a crystal clear, shallow pool, about three inches deep, on the edge of Bluff Creek, northern California. Five toes were clearly imprinted. The rifle round seen in the print is a 30.06 calibre cartridge and measures approximately three and a quarter inches in length.

Photo: P. Byrne.

A natural bridge over Bluff Creek, Six Rivers National Forest, northern California. In the 1960s there were few (if any) proper bridges over wilderness creeks in this region. Natural bridges, such as that seen here were used by the author and his stalwart companions of the Bigfoot Research Project. *Photo: P. Byrne.*

Deer bones scattered around a nine-foot-long bed of moss, deep in the upper watershed of Red Cap Creek, in the Trinity Alps, northern California. The author believed the bed to have been made and used by a bigfoot. *Photo: P. Byrne.*

An artist's depiction for the story of a man named Jack Dover who, in the late 1800s, stated that he encountered and surprised a female bigfoot in the Marble Mountain Wilderness, northern California. The story is contained in a booklet entitled *The Hermit of Siskiyou*, discovered in the Humboldt State Library, Eureka, California by one-time bigfoot researcher Jim McClarin. *Photos: P. Byrne.*

The author's brother, Bryan, points to a hill at Tenmile, Oregon, where in the 1960s two boys claimed to have seen a bigfoot. The next evening a truck driver claimed a similar sighting on a highway, just one mile away. *Photo: P. Byrne.*

(Left) The fast flowing torrent of Bluff Creek, Six Rivers National Forest, in the general area of which, in the 1960s, many sets of large, five-toed footprints were found and attributed to the huge, elusive primate known as bigfoot.

(Right) Small research boat in Bute Inlet, BC, the property of bigfoot hunter Bob Hewes, of Colville, Washington. He and the author used it for bigfoot research in some of the huge inlets of the BC coast. *Photos: P. Byrne.*

The author riding an all-terrain motorcycle called a Tote Goat. These small but tough two-wheelers, which could handle almost any kind of rugged terrain, were a vital part of the equipment of the author's 1960s bigfoot research project. *Photo: P. Byrne.*

(Left) Author (left) camping in the upper watershed of Red Cap Creek, Trinity Alps, northern California, with Jim Crew, nephew of Gerald Crew, a member of the road building gang that made the first discoveries of bigfoot footprints in the Bluff Creek watershed in 1958.

(Right) Weyerhaeuser company surveyor John Hathaway, who, with another company surveyor, Charles Kendall, had an encounter with a bigfoot in June 1978, in the Cascade Mountains, northern Washington. Both men, senior employees of Weyerhaeuser, had decades of outdoor experience in the northwest forests and until that time did not believe in the phenomenon. *Photos: P. Byrne.*

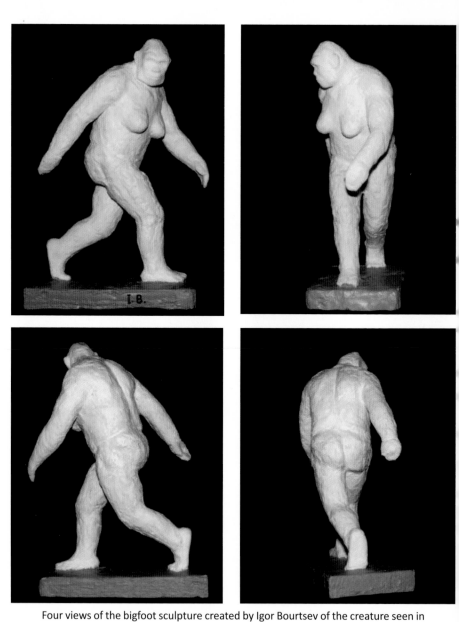

Four views of the bigfoot sculpture created by Igor Bourtsev of the creature seen in the Patterson/Gimlin film. The sculpture is to scale, so provides an insight as to the creature's appearance at different angles. The sculpture was created by Igor in the early 1970s after he and Dmitri Bayanov, both Russian hominologists in Moscow, were provided with a copy of the Patterson/Gimlin film. Both men are convinced that the creature filmed was a real bigfoot.

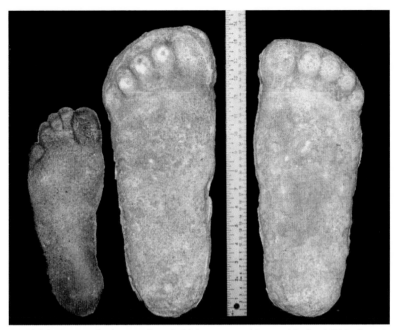

An 11½-inch cast of a human foot (left) is compared here with the 14½-inch casts of footprints left by the creature in the Patterson/Gimlin film. The relative size of the prints (length and width) provides a good insight into the height and weight of the creature. *Photo: C. Murphy.*

Author (left) chatting with Bob Gimlin in 2009. Gimlin, along with Roger Patterson, took the famous 1967 film of a bigfoot at Bluff Creek, California. Bob has attended numerous bigfoot conferences and outings, and never fails to intrigue everyone with accounts of his remarkable experience to impress them with his integrity. *Photo: C. Murphy.*

Author in the map room of his 1970s organization, Bigfoot Research Project II. The project was headquartered at Evans, near Colville, Washington (just south of the British Columbia, Canada border). *Photo: P. Byrne.*

Author (right) with veteran bigfoot researcher Alan Berry (d. 2012) of Sacramento, California, at 8,000 feet in the Dardanelles ranges of the High Sierras mountains of northern California (1970s). *Photo: P. Byrne.*

Bigfoot eyewitness Jack Cochran, of Goldendale, Oregon, stands at the precise place where, in the 1990s, with a companion, he encountered a bigfoot. The creature was watching him from the concealment of a stand of fir trees; when eye contact was made, it immediately turned and walked away. The site of the encounter was at the top of a small, forested peak known as Fir Mountain, which is located on the ridge that forms the upper east side of the Hood River valley, in northern Oregon. *Photo: P. Byrne.*

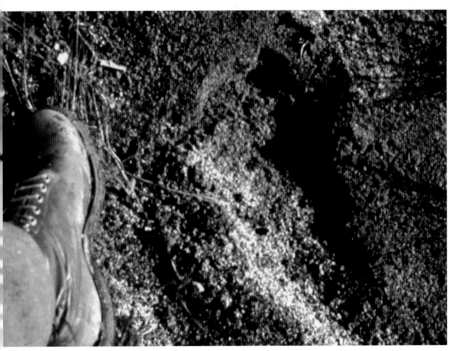

Footprint (one of a series) found in October 1990 in the mountains of the Dark Divide, Washington State, by renowned lepidopterist and author Robert Pile of Grays River, Washington. *Photo: R. Pyle.*

Oliver L. Kirk (Department of Natural Resources Conservation Enforcement Officer, The Confederated Tribes of the Warm Springs Reservation of Oregon) is seen here holding a cast of what was believed to have been a bigfoot footprint. Oliver provided the following report on the cast:

During August 1994 the footprint from which the cast was made was found in the NW area of the Warm Springs Reservation, in an area commonly known as Schoolie. It was brought to my attention by a Bureau of Indian Affairs Fire Management engine crew. They had protected the footprint with stones and small limbs. There were about a half dozen footprints leading up a faint game trail. Most of the prints were indistinct and couldn't be cast in plaster. These tracks followed the game trail leading uphill. The average stride was approximately 32 inches from the tip of the longest toe to the back of the heel. This particular footprint measured 13 inches long and 5½ inches wide. There were more tracks located on another road north of this location at the top of the ridge from these footprints. Previously to this occasion, another engine crew had reportedly observed a small bigfoot creature about a mile further north from this location near a pine tree plantation. The creature was observed walking on a dirt road, then stopping to kick the dusty road.

Author (left) with Jean-Paul Debenat at Bennett Pass, Oregon (near Hood River), July 1995. Dr. Debenat, who lives in France, has traveled to North America several times to research the bigfoot issue. *Photo: J-P Debenat.*

Author (center) with veteran bigfoot researcher René Dahinden (left) and Chris Murphy meeting in 1996, Richmond, BC, to discuss the Patterson/Gimlin film. *Photo: P. Byrne.*

This artwork by Patrick Beaton provides an insight into the massive physical size of bigfoot. This aspect of the creature is often what most impresses eyewitnesses. For certain, great size is a major factor for the survival of many animals in the Pacific Northwest where the climate is cold and damp for most of the year. *Artwork/photo: P. Beaton.*

Generally, most bigfoot sightings are little more that a fleeting glimpse. Patrick Breaton captures such a moment in this artwork. We know bigfoot are extremely shy, and most likely see a person long before being seen. In most cases they leave the scene very quickly when noticed, and this is why it is so important to be able to access and use your camera at a moment's notice.

Artwork/photo: P. Beaton.

A portrait of a yeti by the noted Canadian naturalist artist Robert Bateman. To my knowledge, there is no known reasonable photograph of a yeti. There is perhaps a better excuse for this than for bigfoot because people in the yeti's domain would be much less likely to carry or even own a camera. For this reason, the artist naturally based his painting on written descriptions of the creature, witness drawings, and his own insights. In this connection, Bateman's vast experience with animals would have certainly played a part, so I believe we can give the image a fair degree of credibility. *Artwork/photo: R. Bateman.*

SECTION TWO The Yeti Mystery

CHAPTER 17

The Yeti

One of the last great mysteries of this little planet of ours is that of the yeti or snowmen, the paradoxical creatures of the Himalayan ranges of the country of Nepal. The mystery is contained in two parts: (1) Who, or what, is the yeti? (2) Does it presently exist in remote areas of the inner Himalayan ranges?

As to what the yeti is, apart from being an upright-walking, hair-covered primate of unknown origin, at this time no one knows for sure. Scientific interest in the creature centers around the fact that all evidence points to it being bipedal, or upright-walking. As with bigfoot, what makes this mode of locomotion significant is the fact that the only other primate that moves like this is a human. All of the apes and monkeys (other primates) known in Asia essentially travel on all fours. We will know what the yeti is when we make contact and are able to study the creature. So far no one has been able to do this.

As to their existing in the Nepal Himalaya—evidence of which goes back 400 years in Nepalese and Tibetan legend and lore—if they truly do, how would you go about trying to find one? How would you start? How long should you be prepared to spend? Which areas of the Himalaya would be the most likely to produce a find? What kind of equipment would you take with you and, when you come across one, what do you do?

This book, written by an expert on the subject, gives you the answers. Read it, make your plans based on what it tells you and then head for the most magnificent mountain ranges in the world—land of the smiling Sherpa, the shaggy yak, Buddhist shrines perched on the tops of impossible crags, fifty-foot-high rhododendron trees, gloriously colored pheasants, deer with fangs, wild goats that you can smell a quarter of a mile away, other strange beasties that are half sheep and half goat and, if what the mountain people say is true, big, smelly, hair-covered, manlike creatures walking upright on two legs. The abominable ones. The yeti.

CHAPTER 18

Yeti Hunter

I think you will agree that the title "yeti hunter" deserves some explanation and is a good way to start this book section, especially when the author makes so bold as to offer what looks like his professional advice on how to find a yeti. So, the obvious questions: What is his background in the phenomenon? What experience has he had? What field work has he done, and what does he have to show for it as a result? All of these questions lead up to the most important one: Can he be relied upon to truly give expert advice on solutions to the yeti mystery? Good questions which, in all fairness, the author feels he should probably answer, now, before he goes any further.

I first heard of the yeti from my father, telling me bedtime stories when I was a boy, back in the 1920s. As a result, I actually grew up with some knowledge of the creatures, plus a strong interest in the legend. In those days, of course, the name "yeti" was not commonly known or used. We called them the abominable snowmen. It was not until many years after I first heard about the creatures that I learned of the origin of this marvelous name.

The first Westerner to report the possibility of an unknown, bipedal primate in the Nepal Himalaya was B.H. Hodgson, in 1832. His Sherpa guides, he stated, claimed to have seen a tall, bipedal creature covered in long dark hair. In 1887 he again reported that his Nepalese staff said they had encountered a hair-covered, apelike creature in the Nepal mountains. But it was a British army officer, trekking in the Himalaya in 1889, one Colonel Lawrence Waddell, who really brought the Yei to the attention of the western world. He did this when, after finding mysterious, humanlike footprints in the high Himalayan snows, wrote an attention-grabbing article for the *Calcutta Statesman*. In it, based on a translation of the name his porters gave to the footprint maker, *metah kangmi*, meaning foul-smelling man of the snows, he called the ceature the "abominably smelling man of the snows." From this, in time, came the name, the abominable snowman, one that was soon accepted world wide for Waddell's unknown primate. This, rather than the name yeti, is what was used when I was young and first became fascinated with the mysterious creatures and dreamed of travelling to the Himalaya to find one. However, it was not until I was in my early twenties that I got my first chance to make a personal investigation into the great mystery.

In 1946 I was in Bombay, India, waiting for a ship to take me back to England after four years in the British Air Force, in South east Asia. An Air Force buddy, Peter Saunders, from Harlston, Norfolk, was waiting with me. Having plenty of time on our hands while we waited for the ship, we took a trip to Darjeeling, the old British hill station in the Himalaya. We planned to spend a month there together and do a trek in the Sikkim Himalaya and look for the abominable snowman. Unfortunately, Saunders fell and broke his ankle and had to be sent back to Bombay. I therefore carried on alone.

I spent just two weeks in the mountains, but I was enthralled with the incredible vistas presented by the magnificent ranges, especially those of the Kanchenjunga Himal. This is a magnificent peak—its upper ramparts make it the third highest mountain in the world. Also, hiking up the great Sandakphu ridge that separates India from Nepal, I was intrigued by the endless ranges of hills sweeping westward into Nepal—at that time an unknown country and, for foreigners, forbidden territory.

Before leaving Darjeeling I met with some tea planters who lived and worked there. Intrigued with their life style, I immediately began to investigate the possibility of getting into the tea business in India. After I returned to England, I realized my ambition was back in north India. To make a long story short, two months later (having spent just two weeks at home after four years) I was back in India working for a British tea company. Not only that, but I was stationed on a tea estate from which, on a clear day, the Sikkim Himalaya could be seen. The Himalaya, home of the abominable ones and for me, with their mystery, a challenge that had to be answered.

As a tea planter I was allowed home leave every fifth year. However, we were also allowed one month local vacation time, and as soon as my first year was concluded in 1949, I headed back into the Sikkim Himalaya. This time, for my second trip into the mountains, I had a month to spare. I used it to do a big circuitous search, starting from the capital of Sikkim, Gangtok, going north to Lachen and Lachung, on from there to Green Lake at the base of mighty Kanchenjunga, and then cutting west and coming down the great Nepal–India border ridge through Sandakphu and back to Darjeeling.

I made no finds on that second sortie but, using an interpreter, I picked up quite a lot of yeti lore and learned a great deal about the Himalaya and the people who live in them.

The next time I had an opportunity to get into the Himalaya was in the winter of 1956, by which time I had left the tea industry. Nepal still being a closed country, I went again into Sikkim. This time, however, I hiked a reverse circuit of the previous trip, starting from Darjeeling and going in the opposite direction. Ten days out from Darjeeling, I found some footprints in the scree of a glacier below Kanchenjunga. They were ten inches in length, very broad and with five toes. My two Sherpa porters told me they were yeti footprints and seemed to be quite frightened by the sight of them.

A couple of days later, searching in the northern end of the Rathong Valley, I met with a group of Sherpas from the Darjeeling School of Mountaineering. Among them was Tenzing Norgay of Everest, whom I knew from Darjeeling, and late that night, sitting around a big campfire and trying to keep warm in a bitter wind coming down off the snows of Kanchenjunga, he told me that he had recently met a man in Darjeeling who was very interested in sponsoring a full scale search for the yeti. The man's name was Tom Slick and he had left his address with Tenzing's wife.

Returning to Darjeeling I wrote to Slick, who lived in San Antonio, Texas, and after a few months of correspondence we met in India and then went into the Nepal Himalaya together—the country now being open to foreigners—on a reconnaissance that was to be a prelude to a major expedition. During that reconnaissance we both found footprints at two separate locations and this convinced Slick that a full scale search was warranted. I was to plan it and lead it and the first phase was to be for a year.

The first phase, for which I chose the upper regions of the great Arun Valley of central east Nepal, lasted right through 1957. I came down at Christmas for a week and then went back for a second year (1958), part of which was again spent in the upper Arun and part in the Lang Tang Himal. Returning to the mountains again in January 1959 I spent a third year. The three-year search came to an end in December 1959.

The author's experience and his right to consider himself expert enough in yeti lore to be able to give others advice on how to find one is hopefully evident. In summary: one short reconnaissance in the Sikkim Himalaya in 1949; a first expedition, again in Sikkim, in 1956; and the three-year-long (total) expeditions in 1957, 1958 and 1959. With this as a background then, off we go into the wild blue yonder to look for the abominable ones. We will start by looking at where we are going to go, and how we are going to get there.

CHAPTER 19

The Mighty Himal—Land of the Yeti

The great chain of mountains that is known to the western world as the Himalaya, and to the people who live in them (the Sherpas, Tibetans and Bhutias) as the Himal, and to the Nepalese who abide in the great sprawling mass of hills that lie to the south of them as the Mahalangur Himal, or Mountains of The Great Langurs, begins in southern China and then, changing its name half a dozen times, runs for a thousand miles to the very edge of eastern Europe. On the way it contains hundreds of the highest peaks in the world, many of them over 20,000 feet.

From between these mighty peaks, carved out of the rock and soil through millions of years (mainly by the action of glacial water) run hundreds of deep valleys. These valleys, which begin as jagged ravines at the mouths of the glaciers, run south, and through them rivers rush waters of the great Himalayan glaciers to form a series of other fast flowing rivers. In Nepal the principal of these are, starting in the west, the Sarda, the Tula Beri, the Karnali, the Kali Gandaki, the Sun Kosi and the Arun. There are a hundred other lesser rivers.

The valleys down which these rivers flow allow, in some places, for human habitation and cultivation. However, in their upper regions their sides are mostly too steep for human occupancy, as a result of which they have little or no habitation. In these high areas they are densely forested. The forests are composed of birch, pine, juniper, fir, and rhododendron; their floors are covered with a thick under-brush, much of which is giant fern. This dense growth, beginning in the middle Himalaya, continues as the valleys rise up to 10,000 feet or more. Above this there is more growth, but at this level it begins to thin, with trees giving way to dwarf growth. Nevertheless overall growth continues all the way up to 15,000 feet in places, and even after that, in the form of moss and lichen.

The name "snowman" has given rise, over the years, to the belief that the abode of the yeti is high among the great peaks in the land of eternal snows—but this is not so. It is in the great valleys that sweep down from the peaks that most of the wildlife of the upper Himalaya live, and it is in their depths—in the dark and gloomy forests, among giant trees draped with arboreal orchids and moss, in places where the sun seldom shines, that the yeti most likely have their abode.

A small section of the extensive forests at the base of the Himalaya.

Photo: Image from Google Earth; Image © 2011 GeoEye; Image © 2011 DigitalGlobe; © 2011 Cnes/Spot Image; Image © 2011 TerraMetrics.

CHAPTER 20

Selecting Your Target Area

So, where do you go to find your yeti? Where are they most likely to be found? Do you still believe high up in the frozen wilderness that lies between the soaring peaks? On the tops of impossible 20,000-foot crags, access to which demands sophisticated climbing gear and oxygen backed by support teams? Of course not. In fact very little wildlife lives in those high areas and, indeed, once over the 15,000-foot levels, the most that one will find are the little rodents called mouse hares.

Yeti habitat, the records show, like most of the wildlife of the upper Himalaya, is most likely to be in the deep, thickly forested valleys down which rush the hundreds of mountain streams that drain the waters of the glaciers and feed them into the great mountain rivers as I have discussed. These long, deep, high-sided valleys, with their dense growths of rhododendron, magnolia, spruce, birch and pine, and with ground cover in the form of enormous ferns, harbor a great variety of wildlife.

The list includes bear, leopard, wild goat, two species of goat antelope, red panda, huge langur monkeys (from which the Nepalese names of the mountains is derived), several species of rodents and many birds, including a variety of pheasant.

With their abundant water, food in the form of edible plants, and dense cover, these huge ravines offer perfect habitat for a shy, wild creature like the yeti. Therefore it is to these valleys, and not to the impossible ranges high above them, that anyone searching for the yeti should go. It was in these valleys that we concentrated the prolonged searches that were the thrust of my three long expeditions in the 1950s.

CHAPTER 21

Access to Yeti Country

You probably don't have to be told how to get to Nepal from North America. However, just in case you have not yet been there, here is a little information. From the east coast the quickest way is via Europe. From the west, via Asia. Either way, your destination will be the capital of Nepal—Kathmandu. There are a number of airlines flying out of the US and the best of them is undoubtedly Singapore Air, with its first-class record of fine service and its direct flights into the little kingdom.

Once in Kathmandu the next step is to find out how to get from there to your chosen area, and you can do this by consulting any local, reputable travel agency. There are basically three ways: (1) by walking there, all the way from Kathmandu itself; (2) by road transport as far as you can and then walk; (3) by plane to the airfield nearest to your chosen area and then hike in from there.

You will probably not want to walk all the way. Right? It's a long way from Kathmandu to some of the outlying areas, and as your time will probably be limited you will obviously be looking for the most practical way to get there. Using road transportation (which means either by local bus or using your own hired vehicles) you would, for the far west, go all the way to the mountain town of Baitadi, north of Dandeldura. For the central west you would drive to the hill town of Surkhet. For the middle Himalaya you would aim for the city of Pokhara. Finally, for the Sola Khumbu area, your initial destination would be "roads end," about twenty miles out of Kathmandu at Banepa. However, I will mention that on an annual basis the road out of Banepa, towards Sola Khumbu, is being expanded and now it may be possible to commence one's journey on foot more than thirty or forty miles closer to the mountains.

For the central east Himalaya and the Arun area, you would head for the lowland town of Dharan, via the city and airport of Biratnagar. For the middle eastern mountains, the route is via Janakpur and for the far east, Karkhabeta, on the Indian border, just west of the Bengal Dooars tea districts.

By air, heading for the far west you would fly to Baitadi. For the central west, airfields at Simikot or Jumla. For the middle Himal, Pokhara. For the Sola Khumbu and the Everest base camp area, Lukla. For the middle eastern areas, Biratnagar in the Terai, or Tumlingtar, in the Arun Valley. For the far east, Khakabeta.

Once at your chosen destination (regardless of how you got there) you will now be on foot because at most of these air and transportation access points the roads come to an end. You are therefore unavoidably on foot and ready for some hard hiking into steep and rugged country.

These illustrations show you exactly where you are headed and the "lay of the land" for some of the various towns mentioned in the text. North of the towns lie the snow-covered Himalaya. The habitat of the yeti, as reported from encounters by both local inhabitants—Sherpas and high-mountain Nepalese tribesmen—and a few foreign eye witnesses, is in the thickly forested hill areas immediately south of the great peaks in the 10,000- to 16,000-foot ranges.

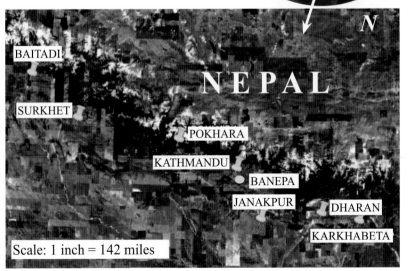

Scale: 1 inch = 142 miles

Photos: Images from Google Earth: (Top) Image © 2011 GeoEye; Data S/O; NOAA; U.S. Navy; NGA; GEBCO; Image IBCAO; © 2011 Cnes/Spot Image. (Lower) © 2011 Cnes/Spot Image; Image © Digital Globe; Image © 2011 GeoEye; Image © 2011 TerraMetrics.

CHAPTER 22

Pinpointing a Search Area

If I were planning a yeti expedition right now, for my search area I would chose either the far western or the far eastern Himalaya of Nepal. My reason for doing this would be two-fold. Firstly, central Nepal, mainly the Sola Khumbu area and the Annapurna regions, has seen a huge influx of people in the form of trekkers within the last few years. The result has been considerable disturbance of the wildlife of the area, enough to cause a shy creature like the yeti to seek less greener pastures, or at least quieter ones. Secondly, indicative of the possibility that the yeti may have been driven out of these central areas is the fact that not one trekker of the thousands now going into the mountains (30,000 a year as of 2000) has reported a yeti sighting or even a footprint find.

At this time far fewer people go to the extreme eastern or extreme western regions, as a result of which they are still relatively undisturbed. In addition, with the exception of short reconnaissances in the nineties by an American group from San Luis Obispo, California, there have been no yeti search expeditions into these areas to date, leaving them, as far as I am concerned, with considerable potential for a find. It might be noted that the American group was organized and led by Kyle Roath who early in 2001 carried out a reconnaissance in the middle and upper Arun of the central east Himalaya. Both areas have deep, forested valleys that provide the kind of cover a yeti would need, as well as the companion wildlife that is found in other areas through the upper middle Himalaya.

A good basic plan for either region would be to fly in to an airport that is central to it and start from there using inquiries in local villages as a means of constructing a sound intelligence base. You would be inquiring mainly for information on sightings and footprint finds and, whenever possible, for information that is as fresh as possible. Nevertheless, when it is credible, old information is good and it should be carefully recorded, no matter how dated it is. One of the reasons for this is that it may relate to more current information and help in the analysis of same. An example of this would be a new sighting in an area that had a long, deep canyon running through it. You would set out to look for signs (footprints) and might find the canyon to be ten miles long. But then your database tells you that there was a sighting there some years before at which time the object seen (a yeti) left the canyon at such and such a point. Using this information you head for that place and start your search there.

Native intelligence, or village information, will be what will determine your search area or areas, and if it provides you with up-to-date findings there is no better place to start than where these findings have occurred.

CHAPTER 23

Permits & Permissions

At this time, as far as is known, one does not need a permit to search for the yeti. However, in the little kingdom of Nepal rules and regulations change quite frequently. If a large expedition is planned, say with six, eight or more people and considerable baggage, then the need for a permit should be given some consideration. If a permit is needed, then I would suggest making application at least six months in advance; Nepalese bureaucracy is not noted for the speed with which it moves.

However, if your group consists of just two or three people and has not attracted the attention of the media, then an ordinary trekking permit covering the area you plan to enter will probably suffice. This can be obtained in Kathmandu from the Nepal Government Immigration Office and you can get directions on how to find it from your travel agent in the city.

I will mention here for your information that a reputable travel agent in the city is Natraj Travel, situated near the University clock tower, just off the Tundikhel (Kathmandu's central park area) in Kathmandu. They will assist with flights to the far east or far west, as well as provide information on trekking permits and/or yeti hunting permits if they are needed. Normal (non-trekking) one to two month visitor's visas are available at the airport on arrival.

CHAPTER 24

Expedition Staff

Your mental picture of a Himalayan expedition heading for the mountains may be of hundreds of porters winding their way through rocky trails loaded with tons of equipment. (Indeed, you may have seen those pictures of the baggage train of the successful 1954 Everest expedition where a thousand porters were used to carry eight tons of equipment to their base camp.) However, unless you plan a major expedition lasting a year or more, with a big team, this will not apply to you. More than likely if there are say, two of you, you will manage with about fifteen men—a team that will be broken down into a sirdar (or foreman), a cook, a general camp man (GCM) and twelve porters.

The sirdar will probably be a Sherpa. He will be fluent in English, so that you can communicate with him and, through him, with your porters and with villagers in the mountains. His job will be to arrange the porter group, choose the cook and the GCM, set the scale of wages to be paid and then run your camp while you are in the mountains.

The cook may be a Sherpa or a Nepalese and it is useful if he speaks English. The GCM may also be a Sherpa, and as he will be the man who will accompany you in your searching, it is important that he also speaks English, but fluently.

The porters will be Nepalese hill men or women. They will carry loads of about 65 pounds each and you can use them all the way from your starting point, which will be your transportation drop-off point, to the snowline.

Your sirdar, GCM and cook will be paid a set wage and will expect to be supplied with food and shelter. Shelter of course means tents. The sirdar may often ask for his own tent while the GCM and the cook will usually share one. (See chapter 26, General Equipment, in regard to renting tents in Kathmandu.)

As mentioned, wages for your porters will be set by your sirdar. Wages may or may not include a food allowance; he will decide that. Your porters will probably not ask for shelter but if you plan to have them carry high, say right up to the snowline, or if you are going to be in the mountains during either of the monsoons, then they may do so. To keep your costs down, instead of renting tents for them, a big tarpaulin (under which they can sleep at night) will usually suffice. Your sirdar will handle any requests they may have in this respect.

Your Nepalese porters will accompany you to the snowline and/ or to your base camp. There they will be paid off and will leave. If you obtained a tarpaulin for them, it makes a very useful shelter for your base camp kitchen.

Your sirdar will arrange for the porters to come back later for the return march out of the mountains, or if you need them before that, say to help relocate your base camp.

A good plan is to keep a couple of your middle-country porters on in camp as full-time staff. Let your sirdar choose them. The reason for having them stay on as extra staff is, in addition to the general work needs of a base camp (such as water carrying, dish washing and firewood supply), you will probably need them to help you with your food supplies (see chapter 28, Food).

If, after you reach/establish your base camp and then decide you want to go higher—above the snowline—you will have to use high-mountain porters. These can be Sherpas or Rais or Tamangs, men of the high mountain tribes. Your sirdar will arrange for these as they are needed and will set a scale of wages, plus agreements concerning food and shelter, if they are requested.

Your high-mountain porters will probably not request clothing, but if you are going high then your sirdar, GCM and cook may do so; this may include boots and socks, trousers and warm jackets. You should discuss this with them before you leave Kathmandu and make arrangements accordingly. Used ex-expedition clothing from any of the many stores in Kathmandu that stocks such is usually quite acceptable.

I will mention here that in my first yeti expeditions in Nepal in the upper Arun valley, in the 1950s, I used Nepalese middle-country porters to carry loads to base camps at 10,000 feet. From there, as I needed them, I used high-mountain, village Sherpas. They were very tough and hardy men, always cheerful and uncomplaining. They walked barefoot on the rocky trails or in the snow all day and the average load was 150 pounds. Their total clothing consisted of a sleeveless jacket of yak wool, a cotton loin cloth and sometimes a wool hat. At night they slept on the snow on beds of juniper branches and moss. Each man carried his own food which was *tsampa* (ground, roasted corn) in a cloth bag, and a pot in which to cook it. Their other personal gear, apart from a basket with a head strap to carry the expedition equipment, was a kukri (a curved Nepalese knife, similar to the machete) in a leather sheath tucked into the front of their loin cloth. A little pouch in the top of the sheath contained a small knife with a flint and some dried moss, to make fire. Those days are gone.

CHAPTER **25**

Travel Logistics

When you march into the mountains from your access point your plan will be to get to your chosen search area and establish a base camp from which you will then conduct your searches. You may move later and over the course of your search periods have several base camps. However, in the beginning you should aim for a single home camp, central to your search area, and your first objective will be to get to this and get it set up.

Thus your first concern in the matter of your expedition logistics will be the time it will take to get to this camp, plus the costs involved. Depending on the number of porters you will use, costs in this area can be a major factor in your budgetary planning.

The person who will work with you on these logistics, and help you plan them, will be a Sherpa sirdar; from experience he will be able to provide you with reasonable, accurate estimates on the amount of time needed to get from your access point—which means your motorized or air drop-off point—to wherever you are going. He will need to have experience because the time needed depends on a number of factors—the principal being the distance covered in your daily march, which may vary greatly.

This variation in distance will be caused by a number of things, including the ruggedness of the country through which you are travelling, the condition of its trails, and the availability of camping space. Surprisingly, in the Himalaya, though there is plenty of space, it is often quite difficult to find a clean, flat area with access to water—the basic needs of even a temporary campsite.

You might note that in the high mountains, flat areas with water access are often used by yak and sheep herders with their animals. As a result, they are often very odorous and dirty, with heaps of animal dung.

Thus, on a march, a suitable camp area may be reached in one day after just a ten-mile march, but the next day only after a long hike of twenty or more miles. Your sirdar will know where suitable campsites are to be found, as will your porters. It is important to keep in mind that it is mainly the availability and location of these campsites that will determine the length of your daily march.

The sorties—or search reconnaissance trips—that you will make from your base camp will of course be different and will usually be one-day hikes only. The logistics, therefore, will be quite simple—out in the morning and back before dark. Take your GCM with you and make sure that you inform your sirdar where you are going and when you plan to be back. Also, equally important, make sure that you have a clearly understood emergency plan in hand with him. The Sherpas are great people, and an experienced and dependable Sherpa sirdar is a marvelous asset to an expedition; but I have found that they work a lot better when they have a clearly understood plan of action to follow, especially in the case of emergencies.

A good rule of thumb for one-day search sorties, where you want to cover as much ground as possible, is to make a note of the time it takes you to get to the furthest distance from your base camp and then give yourself that time, plus a little more to allow for natural fatigue, to get back from there to base before dark. Also keep in mind that whereas time travel estimates like this can be calculated with reasonable accuracy on trails, they may vary greatly if you plan to go off trail. One of the reasons for this is that the off-trail route that you might choose on the way out from camp may, because of brush, or rock formations, or precipitous ascents or descents, be quite difficult to follow again on your return trip.

CHAPTER **26**

General Equipment

The type of equipment you take with you will depend on three things: (1) the season of the year; (2) the altitude in which you plan to operate; (3) the duration of your expedition.

Depending on the time of year, you may have rain. Nepal has two monsoons, one coming into its eastern regions from the southeast in April or May, called the Bay of Bengal monsoon, and the other into the western areas from the southwest in mid June called the Southwest monsoon. Rain can be torrential in these periods and while it may be warm rain in the south (the lowlands), it will be cold and often in the form of snow higher up. Basically, if you are going high, it will be colder. If your search is in lower areas, it will be warmer.

With regard to the duration of your trip, its main effect on your equipment will be wear and tear. All equipment on a Himalayan expedition gets a lot of rough usage; in time even the best of equipment will wear out. Remember this and try to gauge how long you think your gear will last under rugged conditions and rough usage, and plan accordingly.

Start your equipment with a good tent. Don't worry about its weight because it's almost certainly going to be carried by a porter, not by you. (Note: If you are going into the Himalaya for a prolonged period, say three months, then the equipment that you will need to take with you will be too much for you to carry alone. Porters will be necessary and these will be arranged for you by whosoever helps to plan and outfit your expedition in Kathmandu, as discussed in chapter 24, Expedition Staff.)

Make sure your tent is windproof and waterproof, and also that it has a sealed floor with a raised ledge around the outer edge to stop water entering in the event of torrential rain. If you are going for a prolonged period then I would suggest that you make it as large as possible. A well-known expedition equation is that tents get smaller every day in direct proportion to the amount of time they are used and the number of people using them. (Just as backpacks get heavier in direct relationship to the number of hours they are carried and the steepness of the trails up which they are hauled.)

Next, select a good sleeping bag. Most quality bags now come with a temperature rating and if you are going to go anywhere over 15,000 feet then I would suggest that you take along one that will allow for temperatures down to zero. Down bags are still probably the warmest that you can buy in proportion to their weight. But when down gets wet it loses its insulation, whereas a bag insulated with synthetic material will not.

With your sleeping bag you will need a good quality ground pad because at times you may find yourself forced to camp in places where the ground is stony and hard. Inflatable pads are good and can be very comfortable, but remember, a puncture may be difficult or even impossible to repair.

Outdoor clothing is a matter of choice. Rain gear is essential, of course, and it can also double as wind proof gear. Choose basic clothing to keep the body comfortable, keeping in mind that warmth is generated by the body, not by the clothing that covers it. All that clothing does is maintain the heat that the body generates.

Boots, like clothing, are also a matter of choice. Whatever kind you choose, break them in before you leave for yeti country. When purchasing foot gear, don't let anyone tell you that you can hike the high Himalaya in sneakers. That, and I speak from experience, is a good way of acquiring serious foot injuries or even of losing your toes to frostbite.

Lastly, you will be interested to know that Kathmandu now has many stores selling and renting camping equipment of all kinds. The better ones are located in an area called Thamel, not far from the royal palace. Costs for either purchase or rental are reasonable, though in the case of the latter you may be asked for a substantial refundable deposit. You may also be asked to leave your passport as security; do not do this. If returned equipment is damaged or dirty, you may be charged for repair and cleaning. Lastly, if you bring your own equipment from your home country, and after your expedition do not want to take it back with you, these same stores may buy it from you.

CHAPTER **27**

Safety & Specialized Equipment

All good equipment, by which I mean quality equipment, is essentially safety equipment—rain gear that keeps you dry, good clothing that keeps you warm, a sturdy tent that keeps you secure. All are items that contribute to your safety; but after them come some specialized items and they should be discussed.

Probably the principal danger that an expeditioner can encounter in the Himalaya is adverse weather. How does one guard against it? By clothing and by shelter. However, whereas you can wear your clothing and carry additional clothing with you in a backpack, you will probably not be carrying your tent. This, in the event of a sudden change of weather, presents the problem of the need for shelter.

The kukri, the traditional, all-purpose machete-type knife of the Nepalese villager, is an excellent tool with which to make an emergency shelter. Learn how to make a wickiup, the temporary shelter of North American Natives (see sketch). If you have chosen as your search area any of the deep, forested valleys of the north, then all the building materials you will need will be readily available. These will consist of branches for the frame of your shelter and grass or ferns for the roofing. Properly constructed, this crude little shelter will protect you from the elements, particularly from a killer wind with chill factors that could be dangerous.

A North American Native wikiup.
Photo: Public Domain.

You can pick up a good working kukri in Kathmandu for about twenty dollars. Avoid ones with ornamental handles embossed with brass and silver, and make sure that the one you buy has the blade in the form of a continuation of its steel shaft, running all the way back into the handle. Handles should be of wood, but buffalo horn, as long as it is solid, is equally suitable. Use your kukri with care; it will lop off a finger just as quickly as it will cut branches for your emergency shelter.

Continuing in the matter of safety equipment, keep in mind fire-starters. Carry all-weather matches in a waterproof container and make sure they are the kind that will ignite when struck against steel or stone. There are many flammable fire-starter materials now available on the market; choose one that you know or that is tested and guaranteed to work under adverse conditions. An excellent fire builder—once you get even a small fire going—is compressed air. (The canned compressed air normally used for camera cleaning is suitable.)

Lastly, as essentials to your emergency gear, we come to communication and position determining equipment. For the Himalaya there are three important items. These are: a mobile phone, a Global Positioning System instrument (GPS) and an altimeter.

As ordinary mobile, or cell phones, may not work in the mountains, what you have to think about here is a satellite cellular phone. These are, on the whole, quite expensive. However, they are coming down in price. As an alternative to buying one, they can now be leased from major companies like Nokia. An extra battery, for emergency use only, should also be given consideration. The overall cost may be high, but when you consider that in case of an emergency your only way of getting word out (rescue source) may be to a man on foot, it may well be worth it.

On this point you need to know that in the Himalaya, here and there, are police posts. Some, but not all of them, have radio communication with Kathmandu. In the case of an emergency, such as someone down with a broken leg, the usual procedure is to send a runner to the nearest post that has a radio and have the officers there alert Kathmandu. If the officers are able, they will send a message to police headquarters in that city, from where your embassy will be alerted. The embassy will then arrange for an emergency medical evacuation by helicopter. The cost may be considerable, possibly as much as $10,000, and you will be billed for this by your embassy. Note that some medical insurance companies will arrange for coverage for this kind of emergency; standard medical insurance in the US, such as Blue Cross, will not normally cover it. Be aware that your embassy will probably act faster if you have registered with them on arrival in the country, giving them your itinerary and routes.

Next, lets discuss your GPS. There are now several models on the market. Choose a quality item and thoroughly test it and learn to use it at home. Once in the Himalaya use it from time to time to determine your geographical position against your maps. Remember that while your mobile phone can summon help in the event of an emergency, it is via your GPS that you will be able to describe your exact geographical position to a source of assistance. Make sure you include one with your emergency equipment.

Lastly, an altimeter should also be included as part of your essential trio of emergency tools. The principal purpose of an altimeter is, of course, to determine altitude. However it can also be used—aligning it with the contours on your maps—to help determine position, and in the case of a failure on the part of your GPS it can be particularly useful in this respect. For instance, assume you are about half way up a big 12,000-foot hill and the summit is your destination. (Twelve thousand-foot rises in the Himalaya are hills, not mountains.) Because of something like dense forest, or perhaps cloud, preventing you from seeing above or below your position, you are not quite sure how far up the hill you are and what the distance is to the summit. You consult your altimeter and it tells you that you are at 10,500 feet. You look at your map and pick up the 10,500 foot contour line on the hill. Although your altimeter will not tell you exactly where you are on that line, it will inform you that you are on it, and this tells you how far up the hill you are and thus how far it is to the summit. My advice is to include an altimeter as part of your emergency equipment. The Casio watch company makes one that can be worn on the wrist. I have one of these and have personally found it useful and reliable.

CHAPTER 28

Food

There are essentially three ways in which you can supply yourself with food while on a Himalayan expedition: (1) buy it all before starting and carry it all in with you; (2) eat off the country; (3) aim for a combination of both of these plans.

In the interests of logistics and economy, I would suggest number three. This means picking up and taking with you from home and/or from Kathmandu, basic, personally preferred foodstuffs and then using these in combination with what you can find in the mountains to supplement them.

Some time ago the government of Nepal in its wisdom reduced or even eliminated duties and tariffs on imports, including foreign foodstuffs. In the old days it was next to impossible to find even a can of baked beans in the country; nowadays just about every kind of foreign food that you can think of is available in Kathmandu and, to some extent, in some of Nepal's other cities. Swiss chocolate, Danish cookies, English fruit cake, American coffee, cereals, soap, shampoo, toiletry supplies, bottled water…you name it, Kathmandu's stores and little supermarkets now have it.

Basic essentials are those foods which will not be available in the mountains and they will probably include tea and coffee, sugar, liquor, candy, cereals, and other items as a matter of personal taste. What you will find in the hills will be rice, potatoes, vegetables (cabbage, cauliflower, carrots, peas and beans), herbs (such as cilantro, mint, basil), eggplant, tomatoes, eggs, meat (mainly chicken), goat milk, fruit (papaya, pineapple, mandarin oranges, bananas), flour and cornmeal. These, in combination with what you carry in with you, will provide for an adequate and indeed enjoyable diet.

Earlier, I suggested that you keep a couple of your lowland porters in camp for general work. If you decide to go for the combination food supply as suggested here, you will need these men (in addition to doing general camp work) as runners to food source areas to help keep you supplied. You will have to do this because once you get into high country, villages from which you can obtain food may be few and far between. Furthermore, some villages are very small—fifty people or even fewer—with very limited food supplies and, as a result, next to nothing to sell. Hence the need for these extra men.

Work through your cook for this. He will tell you what he wants in the way of supplies and he will provide you with lists. Your sirdar will then tell you approximately what the items will cost and when he does, give him whatever money is needed. He will then give this money, with written or oral instructions, to your two food runners and send them off. They will know where the different items are available and how long it will take them, on foot of course, to reach these sources and carry the foodstuffs back. This is how you will operate the combination food method, and in my opinion it is a good method and one which can keep you supplied with good fresh food for weeks at a time.

CHAPTER **29**

Making Contact

You've been in the mountains three months. You've talked to villagers about the yeti, you searched the beds of a score of mountain streams for signs, you've slept in ice-cold caves and you've survived howling winds and sub-zero temperatures. Your clothes are beginning to fall apart, your boots are cracking, you are tired and footsore, and beginning to think of things like a hot shower, a three-course meal in a fine restaurant, and a big comfortable bed.

Now, it is just after dawn on a fine, clear Himalayan day and you are walking down a long trail that traverses the spine of a 15,000-foot ridge. You are tired and cold and beginning to wonder if it has all been worthwhile—the cost, discomfort, time spent. The blood, sweat and tears.

Suddenly, far down the trail ahead, you see a lone figure walking towards you. The figure, which looks like a man, immediately catches your attention because apart from you, and your men back in camp, there should not be anyone else up here in this back country, certainly not at this time of day. You stop and watch the figure, and as you do you realize that there something odd about its gait—something unusual in the way it is swinging its arms. Something strange about its clothing—all black and reaching up across its neck and into its face. You whip out your binoculars, zoom in on the figure and, as you look at it, a cold chill creeps up your back because you suddenly realize that what you are looking at is not a man. It's a yeti. And it is coming towards you.

As of now it does not appear to have seen you. What do you do? Well, whatever you do you must keep one thing in mind; whatever action you take, now, at this moment of contact, is what is going to decide whether or not you get the prize you want—the record that you are going to take back of the abominable mystery.

Much will depend on your readiness. Have you allowed the long weeks, the cold, the endless searching, to dull the discipline with which you set out, the self training that you imposed upon yourself in this area? Are your recording instruments, your camera and your video, loaded and ready and where you can get at them quickly? We hope so. Because with something as elusive and shy as a yeti you are probably going to get just one chance and this is probably it. And believe me, you had better be ready to take advantage of it because the odds are you are not going to get another. So, what do you do now?

The yeti, although it lives in the high forests, is also a mountain creature and as such probably has keen, long-distance eyesight. Now, if the one approaching has not yet seen you, this is probably because you did the most sensible thing you could do at this moment, which is that the instant you saw it, apart from whipping out your binoculars, you remained perfectly still. Wild creatures always see movement quicker than they will see a still object, even when the still object is something as obvious as a man standing in full sight. So now you can do one of two things:

1. Go for your still camera and camcorder and get them ready to start shooting. If you do, of course, this will necessitate physical action, movement that may catch the eye of the approaching creature and cause it to dive off the trail and disappear into the brush. Keep in mind that if you have a clear view of it, it almost certainly has a clear view of you, even if it has not recognized you for what you are up to this point. So a better plan of action may be as follows:

2. Watch your target until something, a tree or a rock outcrop, obstructs its view of you. The split second this happens, dive for cover. This may mean, if there is no cover available, just dropping down to the ground and getting below eye level. Getting out of the direct view of the approaching creature will allow you to get your instruments ready, get them trained on where you know the creature is next going to appear, and get them ready to shoot without it seeing you do this.

Now you wait for it to again appear, and when it does you do not hesitate to hit it with everything you've got. Use your camcorder to start, rather than your still camera, because if you are using a digital recorder, as is suggested in regard to your equipment, you can later get still pictures from it that may be as good as anything you can get with your still camera. Get your camcorder trained in on your subject and then, if you are lying down as you may be (perhaps by keeping your elbows on the ground), keep your camcorder as still as possible. Use your zoom if you want to and shoot, shoot, shoot. Keep shooting as long as your subject is within view and even afterwards, after it has disappeared; because even though it has gone from your view, it may not have done so from the eye of your camcorder. Indeed, it may still be visible to the eagle eye of your camcorder, walking away through the brush, barely seen, appearing, disappearing. It may also do what many wild creatures do—out of curiosity, stand in concealment and watch you. You won't see it if it does this; but the camcorder may. So keep that finger on the button for another two to three minutes. If you are looking through a viewer, and are a person who shoots with one eye closed, then use the other to watch for any movement.

After your yeti (and remember, this is your very own yeti) has gone, use your binoculars to scan all of the area where it disappeared. It might just appear again on a distant slope, and using your zoom you might be able to get more imagery, every bit of which is part of your grand prize.

Your next step after this—your first contact—is to very carefully remove the video card from your camera and get it into its container, then into a zip lock bag, then into a hard, protective container. Plan to carry it with you, entrusting it to no one. On the question, should you make a DVD (or several) of the card imagery in a city in Nepal? My advice is, if you can access a computer for this purpose, it's a good idea. However, I would not take the card to a dealer and leave it with him.

There has been concern expressed about video cards, DVDs, and computers with regard to x-ray machines at airports. To my knowledge there is no danger in this connection at European, US and Canadian airports. However, Asian airports are a different story and I have had friends who have experienced x-ray damage to their video cards, as well as having films they have taken getting lost, at airports in Nepal. I suggest that you get advice from the people at the Nepal airport you intend to use. Telephone ahead and inquire as to what you should do to protect your card (and DVDs if applicable) if they must be subjected to an x-ray.

Most certainly, don't lose sight of your material (cards, DVDs or films) if you must hand it over for inspection. Make sure you get it back "on the spot." If you have made DVDs, you could of course, put one in your check-in luggage which is not subjected to x-rays. This will at least ensure you have a good copy if anything happens to the card and/or DVDs in your carry-on luggage. The only problem is that you cannot be assured your check-in luggage will not go astray. It is for that reason that you should not put your video card in your check-in luggage.

There's another scenario for your first encounter with a yeti and this might be when you have set up to watch for one from a concealed position, like a blind or hide. Once again your sequence of vital actions will begin when the creature comes into view. First, slow and cautious movement on your part is necessary to get your camcorder aligned, even though the movements you will make are concealed by the covering of your hide. Then, as before, when you are ready, shoot and continue to shoot as long as the creature is visible and for at least another minute after it has disappeared from view with imagery of where you last saw it.

Of the two scenarios, the latter, from the concealment of a blind, is better because, unlike your encounter on a trail or in the forest, you will already have the advantage of cover and not have to seek it with a dive into the brush.

Is there a possibility that you might again see the same creature that you just encountered? There is. It's a slim one if you were in the open when you encountered it because if it saw you (it almost certainly will have seen you), being the wary creature that we know the yeti to be, it will now avoid the area for at least a few days. However, if your contact was made from the concealment of a hide and it did not see you, then a good procedure for a follow-up would be to make a note of the time that you saw it and use the hide again at that time every day for at least a few more days. Wild creatures have set habits. These revolve, for the most part, around their endless search for food, and the yeti that you saw was very possibly where you saw it because it was looking for food there. At the least it might have been passing through the area on its way to another food source. So the possibility of its coming through there again—if of course it did not see you there—is good and worth keeping in mind for the chance of another contact.

Another way to enhance the chance of another encounter would be to stay in the general area for the next few days. If you have decided not to use the hide again then look for some high viewpoints from where there are commanding views of the immediate area. Take a position on one of these, preferably in concealment if this is possible, and use your binoculars to watch and search. If you have a man who will watch with you, take him along; Himalayan mountain men have excellent eyesight and in a watch like this, two pairs of eyes will be better than one. However, do not ask a local man to go with you to your hide and watch with you. Himalayan villagers are good people, but apart from the fact that it is very difficult to get them to keep quiet during a watch, in the event of a yeti appearing it would probably be impossible. The last thing you want when your yeti turns up is to have your companion villager jump up and point and shout in your ear (while the creature is still in the limit of a camcorder range), "LOOK SAHIB. A YETI!!"

Lastly, immediately after the event of your contact, there and then on the spot (or at least immediately you get back to your camp), record in detail everything connected with what happened. Start with what you did when you got up that morning and then go on to say exactly where you went and why. Use your GPS to record the exact position (with altitude) of where you were, where the yeti was, where it came from, and where it went. Describe the weather, including wind and velocity. Describe the vegetation and the wildlife—the other wildlife you know to be in that area. Provide information on your companions, if applicable, with contact numbers and addresses. If you have companions back in camp, get them to record and certify the time you left camp and the time you returned. Include your feelings both at the time of the encounter and afterwards. Make sure you keep this record in a safe place and make a copy of it as soon as you can, and then keep the copy in a safe place. These requirements are necessary because a year from now, ten years, twenty years, you will have forgotten most of the details (certainly the minor ones), and this, your written record of all that happened, will be of vital importance for both you and posterity—and most importantly the credibility of your achievement.

CHAPTER 30

The Evidence

Judging by the creature's footprints, the yeti is likely very different from humans and other hominids. The configuration of the toes appears to indicate that we are dealing with something that is highly apelike. Given the footprints found—which are around 12 inches long—are somewhat representative, then the height of the creature would fall between six and six and one-half feet. Although this is not exceptionally tall by North American standards, people in the region are generally well below six feet tall, so this would possibly give rise to exaggerated claims of the yeti's height.

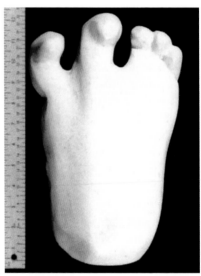

Cast copy of a possible yeti footprint found in 1951. The original was made using a photograph of a print in snow. *Photo: C. Murphy.*

At one time in a temple in the village of Pangboche, high in the Sola Khumbu district of north central Nepal, there was a skeletal hand said to have been that of a yeti. I personally examined this hand in 1959 and obtained from it the complete thumb and the index proximal phalanx bone for scientific evaluation in Great Britain. I replaced the bones taken with human bones, leaving the hand intact. The initial scientific tests on the bones, along with photographs of the entire hand, indicated that the yeti bones were human. Subsequent examinations also indicated, in general, that the hand was human, although possible animal bone content was mentioned in one report. This was long before the advent of DNA processes. Unfortunately, the actual hand was stolen from the temple in the 1980s. Just what became of the bones I provided for analysis was not known to me until 2011 when they were discovered by Matthew Hill, a young BBC reporter, in a laboratory at the Royal College of Surgeons in London, England. DNA analysis was performed and it was concluded that the bones were human.

Nevertheless, I will say that the hand was intriguing. It was certainly not larger than a small man's hand or a woman's hand (Asian heights), so, if real, it would indicate that the yeti in this case was not beyond human standards in height.

Furthermore, there are two known alleged yeti scalps in Nepal. One is in the Pangboche temple, and the other in a temple in Kumjung village (near Namche Bazaar). This second scalp was sent to Chicago in 1960 for scientific analysis which indicated it was made from the skin of a serow (a goat-antelope of the bovine family). Later I was told the scalp actually had been made from serow hide by the temple custodians because of resentment for all the public attention (and possible donations) the Pangboche temple lamas were receiving for the scalp in their possession. Whatever the case, we can speculate that the tall conical shape of the scalp at least represents that of a yeti, and thereby gives us more insight as to what the creature might look like.

Skeletal hand said to be that of a yeti. *Photo: P. Byrne.*

The alleged yeti scalp at Pangboche temple in the Sola Khumbu region of north central Nepal. According to the author's records and in regard to incorrect international media reports of its scientific examination and analysis, this artifact has never left its place of keeping in the temple of Pangboche. *Photo: P. Byrne.*

We know, of course, that the creature is hair-covered. Its hair is likely quite long for added protection. Color ranges from off-white to reddish. Females are said to have large, sagging breasts. A very early illustration of the creature shows that it sleeps in prone position (see page 130). I can only assume that someone apparently saw one sleeping.

The actual facial features of the yeti are said to be apelike. The painting of the creature by the noted Canadian wildlife artist, Robert Bateman (see page 76) is probably quite accurate. However, that a creature of this nature would have feet like those indicated by its footprints requires some thought. Evidently it evolved with unusual feet to facilitate movement in its very rugged and varied environment.

The question as to why the yeti goes above the snowline has been debated. As I have mentioned, its normal habitat is likely in the green belt below the snow where there are rich and varied food sources. Why then would it venture into the high snow-covered areas? A possible explanation for this was offered by two professionals, Don Abbott and Frank Beebe of the Royal British Columbia Museum in Victoria. They pointed out that in North America wolverines go up into the snow to bury meat, where it freezes. They later retrieve their stash when food gets scarce at lower levels. They take it to a lower level, let it thaw, and eat what is naturally just as good a fresh kill. It has been reasoned that both the bigfoot and the yeti might do the same. For certain, the process would work equally well with any solid food item (meat, vegetables or fruit).

Here then are some possible pointers as to the subject of your search. Essentially a bipedal creature about six to six and a half feet tall, covered in fairly long hair (different white to reddish shades), having a possible cone-shaped head, and leaving five-toed footprints of unusual configuration. I am told that it howls and has been reported to kill yaks. To my knowledge, it does not have a distinctive powerful odor like sasquatch or bigfoot. Nevertheless, all animals have an odor, so anything unusual that is sensed should be noted.* As the creature might bury food stuffs in snow, recently disturbed snow patches should be checked. If anything were found buried, my advice would be to leave it and stake out the area (naturally, taking photographs of what was found).

*It should also be kept in mind that the original name for the yeti, given by a Sherpa to the British explorer Colonel Waddell who in 1889 found and reported the first footprints, was *meta kang mi*, which Waddell translated as "foul-smelling man of the snows," and this in turn gave rise to the popular name "abominable snowman."

CHAPTER 31

Summing Up

That's it then. You've got the evidence you came for and it's in the bag, so to speak. Now, what do you do with it? Will you donate it to science? Will you try and sell it to the media? Will you use it as the basis of a television documentary on the mystery of the abominables? Whatever you do, just remember that you now have, with your video, what may well be historic evidence of the existence of something hitherto completely unknown to science. Keep this in mind and guard your recording carefully. The minute you reach civilization download your video to your computer and make several DVDs. Do not erase your video card. Put it back in its plastic container and store it in a safe place. Do not let anyone persuade you to loan them the card unless absolutely necessary (perhaps a university or scientific institution), and if this is the case, then wherever it goes, whatever is done with or to it, stay with it to personally supervise the work. In other words, don't let it out of your sight.

On this last point, I think it is important to relate the fate of the last major piece of bigfoot film evidence discussed in chapter 14— the 16mm film footage taken by Roger Patterson and Bob Gimlin in Northern California in 1967. The master (original) film roll was borrowed by a film maker for the purpose of making a documentary. In this process, the film was lost and to this day, despite exhaustive searches by many people, including this author, it has not been found. Fortunately, copies had been made of the film before it was provided to the film maker. However, access to the original film is required by scientists and researchers, and because the film is missing, many questions are left unanswered. Although film technology of today is far different to what it was in 1967, without doubt there will be scientific reasons to get access to a video card that contains the images of a yeti.

If, when you get home, you release information about your achievement to the media, you will undoubtedly receive a lot of attention. Use this attention prudently and with great care, keeping in mind that everything you say in connection with your video, and everything you do will contribute to its credibility in the mind of the media and, through them, the public. It can also be used to discredit the credibility of your achievement. An unpleasant prospect? Yes. However nowadays, when you present something new to science, and to the media, especially when you do not have physical evidence to back it up, you had better be very careful about how you do it. This is one reason why, if you have followed the suggestion I made earlier, you will have recorded, in detail, all of the events surrounding the personal contact you had with your very own yeti.

So, are you getting ready to start out? Preparing to head for the highest mountains in the world to search for the most elusive of the world's mysterious creatures, the abominable ones? Are you looking at airline timetables and wondering how much it is going to cost, who you are going to take with you, how long you can put aside for search among the soaring peaks of the world's greatest mountains? Are you thinking about what it will be like to actually encounter one of the creatures, alone in the depths of one of the great forests of the inner Himalaya, or out on the edge of some far-flung crag, with the wind moaning in the rocks and the nearest human beings only your own men, miles away back in camp? Can you already feel that cold chill that is going to creep up your back, sense the hair beginning to prickle on the nape of your neck? Yes? Yes…because if you do make contact with a yeti, that is what is going to happen to you. That, followed immediately by the feeling of triumph you will have in the knowledge of what you have been able to achieve. An achievement that many others have considered impossible to the point where they have given up before they even got started. Will you do that? Of course not. I am going to go for it, you will say to yourself. I am going to give it my best shot, and if I see my associates raising their eyebrows and hear them whispering with each other behind my back, I won't care. I will ignore them and silently, to myself, I will shape Tennyson's immortal lines—to strive, to seek, to find and not to yield—to my plans. I will enforce their meaning by remembering what Sir Edmund Hillary said after his great climb of Everest with Tenzing Norgay, "Nothing ventured, nothing gained!"

Going after the yeti is more than just an expedition. It is a magnificent adventure, out of which will come, for those who dare to take it on, regardless of the consequences, nothing less than a singular and unique sense of achievement, a sense that a challenge was received and accepted.

I'll be looking for your book. I am sure it will be a best seller, and very rewarding to me will it be if you include in it a little mention to your readers that this guide book—the scribblings of an old Himalayan hand—were what stirred you on to success.

CHAPTER 32

Yeti-related Incidents Summary

Early Belief

According to H. Siiger, the yeti was a part of the pre-Buddhist beliefs of several Himalayan people. He was told that the Lepcha people worshiped a "Glacier Being" as a God of the Hunt. He also reported that followers of the Bön religion once believed the blood of the *mi rgod* or "wild man" had use in certain mystical ceremonies. The being was depicted as an apelike creature who carries a large stone as a weapon and makes a whistling sound.

Nineteenth Century

1832: James Prinsep's *Journal of the Asiatic Society of Bengal* published trekker B.H. Hodgson's account of his experiences in northern Nepal. His local guides spotted a tall, bipedal creature covered with long, dark hair, which seemed to flee in fear. Hodgson concluded it was an orangutan.

1887: English explorer B.H. Hughes reported that his Nepalese staff saw a hairy apelike creature in the high mountains.

1889: Colonel Laurence Waddell found mysterious humanlike footprints in the Himalayas. He reported the finding in his book *Among the Himalayas* (1899). Waddell provided his guide's description of a large apelike creature that left the prints, which Waddell thought were made by a bear. Waddell had heard stories of bipedal, apelike creatures but wrote, "none, however, of the many Tibetans I have interrogated on this subject could ever give me an authentic case. On the most superficial investigation it always resolved into something that somebody heard tell of…"

Twentieth Century

1906: H.J. Elwes, botanist and explorer, reported seeing a yeti run over a ridge.

1915: J.R. Gent, British Forestry Officer, Darjeeling Divisor, Phalut area, India, reported finding yeti tracks. Local inhabitants stated that they had seen the creature.

1920: Hugh Knight, a British explorer, stated he came face to face with a yeti carrying a crude bow.

1921: Lt. Col. C.K. Howard-Bury found footprints on Mount Everest at the 21,000-foot level.

1925: N.A. Tombazi, a photographer and member of the Royal Geographical Society, wrote that he and others in a group saw an unusual creature at about 15,000 feet near Zemu Glacier. Tombazi later wrote that he observed the creature from about 200 to 300 yards for about a minute. He stated, "Unquestionably, the figure in outline was exactly like a human being, walking upright and stopping occasionally to pull at some dwarf rhododendron bushes. It showed up dark against the snow, and as far as I could make out, wore no clothes." About two hours later, Tombazi and his companions descended the mountain and saw the creature's footprints. He described them as, "Similar in shape to those of a man, but only six to seven inches long by four inches wide." He went on to state, "The prints were undoubtedly those of a biped." Tombazi was unable to get a photograph of the oddity.

1931: Wing Commander E. Bentley Beauman, Royal Air Force, reported that he found yeti tracks at the headwaters of the Ganges River, India.

1935: Villagers at Kathagsue reportedly drove off a yeti after it had killed two sheep.

1936: Eric E. Shipton, mountaineer, reported that he found yeti tracks on his way back to Kathmandu from Everest.

- Ronald Kaulbach, botanist and geographer, reported that he found yeti tracks at the 16,000-foot level on a pass between the Chu and Salween Rivers near Bumthang Gompa, Nepal.*

1937: A British traveler reported that he found yeti tracks on the Biafua Glacier.

- F.S. Smythe reported that he found unusual tracks in the Bhyundar Valley, Garwhal, India. They were said to have been made by a bear, but there was some reasonable doubt.
- The first photographs of footprints are taken by F.S. Smythe in a 16,500-foot pass in central Nepal.

1938: A British monument curator from Calcutta reported being kidnapped by a yeti while vacationing in Sikkim. He claims the yeti took him to a cave and later released him.

1939: Locals in the Himalayas were reported to have got a yeti drunk by leaving liquor at a wellhead. The creature was captured and bound, but later sobered up, burst his binding and ran off.

1940: Slavomir Rawicz claimed in his book *The Long Walk* (published in 1956) that while he and others were crossing the Himalayas in the winter of 1940, their path was blocked for hours by two bipedal yeti-like creatures. Rawicz and the others simply observed the creatures until they went away. In 2009, Rawicz was accused of stealing the story of his "long walk" and fabricating the sighting account provided here.

1944 (circa): An engine driver of a goods train reported that he saw a giant figure ambling along the railway track just before reaching Siliguri, West Bengal, India. That same night, a woman reported that a giant yeti had come to her door, whereupon she immediately shut the door, screamed and fainted with fright. The next morning footprints eight inches wide and fourteen to fifteen inches long were found in the soft mud at the location. Geshe Chomphel, a Mongolian lama who happened to be in Siliguri, made exact tracings of the prints. Unfortunately they have not come to light.

* A careful examination of the information provided by Kaulbach indicates that he was in complete error concerning the whereabouts of his find. There is no Bumthang Gompa in Nepal, and no Chu River. As to the Salween River, its origin is in China, with its nearest point to Nepal several hundred miles away to the northeast in China.

1947: A yak breeder named Dakhu, a resident of Pangboche, stated he saw a yeti at a distance of fifty yards. The creature simply walked away.

1948: A lama, reported that he saw a yeti after a heavy snowfall. He said he heard it screaming and saw it come down from the slopes walking on all fours. It then stood up on its hind legs and scratched its chest. He said that in general it was a rather stunted animal, hair-covered with a conical head.

▪ Peter Byrne reported finding a yeti footprint in northern Sikkim, India near the Zemu Glacier, while on holiday from a Royal Air Force assignment in India.

1949: Sen Tensing, a native of Nepal, reported that he and a number of other men saw a yeti at Thyangboche at a distance of about twenty-five yards. He described it as half-man, half-beast, standing about five feet, six inches, with a tall, pointed head. Its body was covered with reddish-brown hair, but it did not have hair on its face.

▪ A Pangboche villager named Mingma stated he heard yells and reported that he saw a yeti. He took refuge in a stone hut and observed the creature.

▪ It was reported that a yeti came out of the forest and played about near the Thyangboche Monastery. It was driven away by the lamas beating gongs and blowing trumpets.

1950: A Sherpa named Sen Tensing, in the company of others, stated the group saw a yeti at about twenty-five paces near Thyangboche.

▪ A lama named Tsangi (or Tsangyi) at the Thyangboche Monastery said that all the lamas there have seen a yeti. He stated, "It was early evening when we saw it sitting well up the slope of Ka Taga. It was a huge creature with yellowish long hair growing downward from the waist and upwards on the upper part of the body. There was no hair on the face, but its hair, which was short on the top of the head, grew to a point. All of the lamas blew horns and shouted and it bounded away up the mountain leaping with its back legs together and front legs, or arms together. The back feet faced backwards."

1951: While attempting to scale Mount Everest, Eric Shipton took photographs of a number of large footprints in the snow at about

20,000 feet above sea level. Shipton declared that although he was previously skeptical about the yeti, he now believes in the existence of the creatures.

1952: Sherpa Pasant Nyrima and others went to look for a yeti and reported that they saw one near Namche Bazar at a distance of 200 yards.

■ A Thammi villager by the name of Ang Shering and his wife reported seeing a yeti near a forest.

■ Mrs. Mira Bahen, who works at the Gopad Shram (Hermitage), Himalayan Tehri Gerhwal District, Northern Uttar Predish, India, reported that villagers there say they have seen yeti footprints in their fields, and have told of actually sighting the creature. One old villager told her, "My buffalo had been troublesome about giving milk in the daylight and one night I went to the shed where it was standing. I went up to the buffalo to start milking when suddenly I saw a terrifying figure moving along the mountain side about twenty yards above me. I was so alarmed that I crouched down and peeped out from behind my buffalo. Oh, it was very tall with great hairy shoulders as if it was wearing a big fur coat. It moved in an upright position, just like a huge man. It did not leap or hop. It went striding along, but I could not see exactly how its legs worked. It never looked my way and passed off along the mountain." Another villager said he saw a similar creature with "a little one playing around it."

1953: Sir Edmund Hillary and Tenzing Norgay reported seeing large footprints while scaling Mount Everest.

■ Tibetan lama, Tsultung Zangbu, reported that while traveling in Assam he met a yeti which was carrying two large rocks. It simply passed by.

■ Edmund Hillary reported that he found possible yeti tracks in the Barum Khola range.

1954: The *Daily Mail* reported (March 19, 1954) on the analysis of hair samples from an alleged yeti scalp found in the Pangboche temple. The hairs were black to dark brown in color in dim light, and fox red in sunlight. The hair was analyzed by Professor Frederic Wood Jones, an expert in human and comparative anatomy. During the study, the hairs were bleached, cut into sections and analyzed microscopically. The research consisted of taking microphotographs of the hairs and comparing them with hairs from known animals such as bears and orangutans. Jones concluded that the hairs were not actually from a scalp. He contended that while some animals do have a ridge of hair extending from the pate to the back, no animals have a ridge (as in the Pangboche scalp) running from the base of the forehead across the pate and ending at the nape of the neck. Jones was unable to pinpoint exactly the animal from which the Pangboche hairs were taken. He was, however, convinced that the hairs were not of a bear or anthropoid ape. He suggested that the hairs were from the shoulder of a coarse-haired, hoofed animal.

▪ *Daily Mail* Snowman Expedition mountaineering leader John Angelo Jackson tracked and photographed many footprints in the snow, most of which were identifiable. However, there were many large footprints which could not be identified. These flattened footprint-like indentations were attributed to erosion and subsequent widening of the original footprint by wind and particles.

▪ Two members of Edmund Hillary's expedition reported possible yeti tracks were found in the Choyang Valley.

— The head of a Swiss expedition reported that the group had found possible yeti tracks. They were photographed by expedition member, Norman G. Dyhrenfurth.

▪ Lamas of the Thyangboche Monastery stated that in the winter yeti are seen near the monastery playing in the snow. The lamas are all firm believers in the creature's existence. We are told they kill yaks, skin them carefully, and plant the horns in the ground. We are also told that a Sherpa ran and locked himself in his hut when he saw a yeti running towards him. When the Sherpa emerged from the hut, he found his yaks lying dead in a pool of blood. The yeti is generally described as not more than 5 feet in height, with reddish-brown hair all over its body, and a conical head.

▪ It was discovered in an old map of the Everest region, based on surveys made in 1921, that the Everest mountain range was called Mahalangur Himal, which means "Snow Mountains of the Great Apes." Although apes are well known in Nepal and Tibet, the natives never speak of the yeti as an ape. They speak only of yeti, kanag-mis, mirkas, or mi-gos. A Sherpa named Tensing, and other Sherpas, stated to a correspondent they believe there are two varieties of yeti. One attacks yaks; the other attacks humans.

1955: Members of an Argentinian mountaineering expedition led by Huerta, reported that one of their porters was killed by a yeti.

▪ Members of a Royal Air Force Mountaineering Expedition reported finding yeti tracks.

1956: John Keel reported that he followed yeti tracks for two days and finally saw one of the creatures in a swamp.

▪ Natives in Kathmandu reported that the carcass of a yeti had been spotted. They said that it rests in an icy crevasse at the base of 27,790-foot Mt. Makalu, located on Nepal's border with Tibet. A British tea planter, an avid mountaineer, said he might lead an expedition this year to verify the find.

1957: Members of Tom Slick's expedition reported that they found tracks in two different locations in the upper Arun Khola valley.

▪ Traces of footprints reminiscent of yeti prints found in the Himalayas were reported found in the Pamir Mountains tableland and on the Khingan Mountains in Russia. Soviet researchers are currently comparing the prints.

▪ Soviet scientist Professor Alexander G. Pronin, leader of a Leningrad University expedition to find the abominable snowman, stated that he sighted the stooped, hairy creature twice (August 10 and 13, 1957) looking down on him from an icy peak on a plateau of Fedchenko Glacier (6,000 feet high). He refused to call it human, pointing out that it wore no clothes in that harsh climate. "I can only say what I saw," Pronin said. "I can make no categorical claim." He said he saw the creature at a distance of 140 yards, and that it walked on two legs, was stocky, covered with reddish fur, and definitely was not a monkey or "simian type." He went on to point out that local inhabitants, who believe in the creature, reported the disappearance of pots and pans, even laundry. Sometimes the laundry reappeared weeks later flapping from some crag or cliff. Several natives described the creature as mischievous. Pronin thinks that there may be connection here regarding the disappearance of his expedition's rubber boat.

1958: A Mount Everest expedition, headed by British mountaineers Erick Shipton and Michael Ward, found unusual footprints apparently made by a bipedal creature, on the slopes of Menlungtse at about the 19,000-foot level. The prints proceeded for about a mile and then became lost in a moraine of ice. Shipton stated that the prints were "slightly longer and a good deal broader than those made by our large mountain boots." He said that the tracks were very fresh, probably no more than twenty-four hours old. He went on to remark, "Where the tracks crossed a crevasse, one could see quite clearly where the creature had jumped and used its toes to secure purchase on the snow on the other side."

Norman G. Dyhrenfurth reported that a "reliable" native Sherpa guide, Dava Temba, saw a yeti. Temba said he saw a four-foot yeti collecting frogs in a rivulet at night, and the creature chased him when he caught it in a flashlight beam. "Dava Temba ran to inform us," the professor said, "but we found the yeti gone by the time we had marched one mile to the spot."

Dyhrenfurth further reported that he visited caves in which the yeti lived and collected droppings, hair, and enough other proof to convince scientists of its presence. "Our investigations revealed they [yeti] were of two varieties, one about 10 to 12 feet tall, and the other smaller—about four to five feet high," he added.

"While our investigations are still unfinished," Dhyrenfurth stated, "we want to declare that the yeti is no more a myth—but hard truth. We return fully convinced the yeti is a humanlike, rare and fast-disappearing creature possessing the intelligence of a normal grown-up man."

Riflemen of the Raja of Mustang (Kingdom of Mustang, India) reported that they pursued and killed a creature that from its description had perhaps some relationship to the allusive yeti, although only four and one-half feet tall. It was spotted in a 14,000-foot high mountain pass near the border of northwestern Nepal and Tibet, and was said to have run nearly a mile on two legs carrying a yak the size of a full-grown cow. "The animal was cornered in the pass with its prey," one of the riflemen said, "We tried to crush him by throwing down boulders, but we only injured him. He made a strange whistling sound when he was hurt. He tried to get away, but we finally shot him down." When the hunters had a good look at the beast, they were shocked, "Its face was like a bear's," the spokesman said, "but its feet were human." It was covered with long, fine hair, less coarse than a bear's.

1959: A Japanese expedition to find the yeti under Fukuoka Daigaka reported that tracks were found.

- An expedition under Professor T. Ogawa reported finding tracks.
- Members of the third Slick Expedition found and followed yeti tracks. Also, supposed yeti feces were collected. Analysis found a parasite which could not be classified. Cryptozoologist Bernard Heuvelmans wrote, "Since each animal has its own parasites, this indicated that the host animal is equally an unknown animal."
- Bones from an alleged yeti hand are analyzed in Britain and reported to be human.

1960: Edmund Hillary sent a supposed entire yeti scalp from the Khumjung Monastery to the West of Nepal for testing in Britain. The results indicated the scalp was manufactured from the skin of a serow, a member of the goat-antelope family. Anthropologist Myra Shackley disagreed with this conclusion on the grounds that the hairs from the scalp look distinctly monkeylike and that it contains parasitic mites of a species different from that recovered from the serow.

1964: Peter Taylor, an Australian, found footprints 20,000 feet above Lang Tang Valley.

1969: A villager in Tibet reported seeing a yeti, which he described as a reddish-brown figure that stood up and looked like a super-human. Chinese troops with guns, ropes and nets went to search for the creature, but there was no indication that it had been captured.

1970: British mountaineer Don Whillans claimed to have witnessed a creature when scaling in the Annapurna region. According to Whillans, while scouting for a campsite he heard some odd cries which his Sherpa guide attributed to a yeti's call. That night, he saw a dark shape moving near his camp. The next day, he observed a few humanlike footprints in the snow, and that evening, viewed with binoculars a bipedal, apelike creature for twenty minutes as it apparently searched for food not far from his camp. He described the creature as "something between a gorilla and a bear."

1972: Two Americans, zoologist, Edward Cronin and mammalogist Jeffrey McNeely, made plaster casts of footprints.

1973: British mountaineer Sir John Hunt found new footprints.

1974: Members of a Polish expedition to Mt. Lhotse photographed footprints.

A 19-year-old girl claimed she was attacked by a yeti.

1975: A young Polish trekker, Janus Tomaszczuk, claimed to have been approached by a yeti near Mount Everest.

1978: A Sikkimese forest ranger reported an attack by a yeti.

Soviet news agency Tass reported yeti-like creatures sighted in northern Siberia near Yakut.

Lord Hunt photographed yeti tracks.

1984: Famed mountaineer David P. Sheppard of Hoboken, New Jersey, said he was followed by a large, furry "man" over the course of several days while he was near the southern col (gap between peaks in a mountain range, used as a pass) of Everest. His sherpas, however, say they saw no such thing. Sheppard took a photograph of the creature. Later study of the photo proved inconclusive.

1986: Climber Reinhold Messner reported a close-up sighting of a yeti as it came into sight from behind a tree.

1992: Julian Freeman-Atkwood and two other men camping at a secluded spong on a remote glacier in Mongolia reported finding an unusual trail of heavy footprints one morning in the snow outside their tent. They said the prints were definitely made by a creature larger and heavier than a human.

1998: American climber Craig Calonica, reported seeing a pair of yetis on Mount Everest while coming down the mountain on its Chinese side. Both creatures had thick, shiny black fur, he said, and walked upright.

Twenty-first Century

2001: A hair sample said to be from a yeti was collected in Bhutan. The hair was later analyzed in Britain by Bryan Sykes, Professor of Human Genetics at the Oxford Institute of Molecular Medicine. Sykes stated, "We found some DNA on it, but we don't know what it is. It's not human, not a bear nor anything else we have so far been able to identify. It's a mystery and I never thought this would end in a mystery. We have never encountered DNA that we couldn't recognize before." **However, in 2012 Dr. Sykes reported that the hair was subsequently found to have originated from a bear. He is currently conducting research on other alleged hominid hair.**

2002: Tribal villagers in the northeast Indian state of Meghalaya claimed to have seen a strange manlike creature. Mr. Nebilson Sangma revealed that he had stumbled across a scary, furry creature on a hunting trip in the jungles of the West Garo Hills. "After overcoming the initial shock, my brother and I observed this gigantic hairy creature for three consecutive days from afar," he said. Some time later, Mr. Dipu Marak captured on video the nesting place of the creature, which apparently revealed telltale signs of the existence of Mande Burung—the mythical monster in these parts of India, akin to bigfoot or yeti of the Himalayas. Wildlife experts and biologists then discovered footprints measuring some fifty centimeters in the same area—the foothills of a peak in the Nokrek National Park.

2003: Yoshiteru Takahashi, leader of a Japanese expedition, claimed to have observed a yeti

2007: American television presenter Joshua Gates and his team (*Destination Truth*) reported finding a series of footprints that resembled yeti prints in the Everest region of Nepal. Each of the prints measured thirteen inches in length with five toes that measured a total of 9.8 inches across. Casts were made of the prints for further research. The prints were subsequently analyzed by Dr. Jeffrey Meldrum, a leading evolutionary morphologist, and preliminary data suggests that the footprints are anatomically legitimate and do not belong to a known primate.

2008: A team of seven Japanese adventurers photographed footprints which could have been made by a yeti.

2009: Joshua Gates and his team (*Destination Truth*) led another expedition to the Himalayas. They found a hair sample that did not appear to match any known species of animal. They also found an animal limb that had clearly been torn straight off an animal. They suspected that this was the work of a yeti. The hair was DNA tested and found be from a non-human primate.

2011: DNA analysis on the hand bones sent to Great Britain in 1959 confirmed that they were human (as discussed on page 112).

References: Wikipedia – Creative Commons>http://creativecommons.org/licenses/by-sa/3.0/< and Ivan T. Sanderson, 1961. "Abominable Snowman: Legend Come to Life," and author's general files.

Yeti Photo Album

Photographs presented in this section illustrate the author's expeditions in search of the yeti. One will immediately see the grand ruggedness of the Himalaya region and thereby fully appreciate the author's words on being properly prepared for yeti field research.

Footprint found by Tom Slick in the upper Iswa Khola—Iswa River gorge, elevation 10,000 feet—during the 1957 reconnaissance. The footprint length was about 12 inches and there were five toe impressions. *Photo: P. Byrne.*

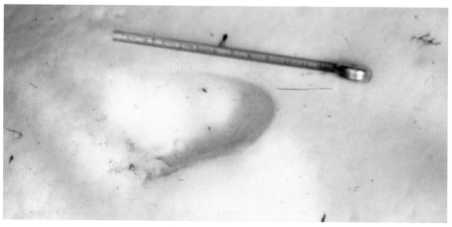

Footprint in snow found by the author in the upper Choyang Khola region at 15,000 feet during the 1957 reconnaissance. Toe impressions were faintly visible. *Photo: P. Byrne.*

A Pangboche temple custodian holding an alleged yeti skeletal hand discovered at the temple by the author in 1958. The hand was later stolen (possibly by a foreign tourist) and to this day has not been recovered. *Photo: P. Byrne.*

One of the lama custodians of Pangboche temple wearing the famous "yeti scalp." It is still kept in the temple today and may be viewed and photographed by visitors for a small fee. Note: As can be seen, it is a scalp only, not a skull as often reported. *Photo: P. Byrne.*

Close-up of the yeti scalp on display at Pangboche temple. Its reddish colored hair is stiff and bristlelike and definitely resembles the hide of a serow (a Himalayan goat-antelope), as was determined for the scalp obtained from the temple at Kumjung village.
Photo: P. Byrne.

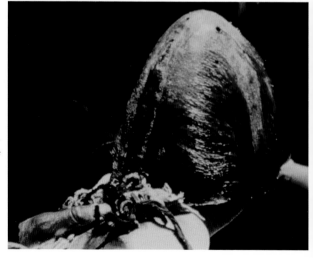

128

A Sherpa man, the father of Lama Yeshi, the Rimpoche (head lama) of Thyangboche Monastery,
showing the author two pony saddle bags made from the skin of the Himalayan goat-antelope,
the serow. *Photo: P. Byrne.*

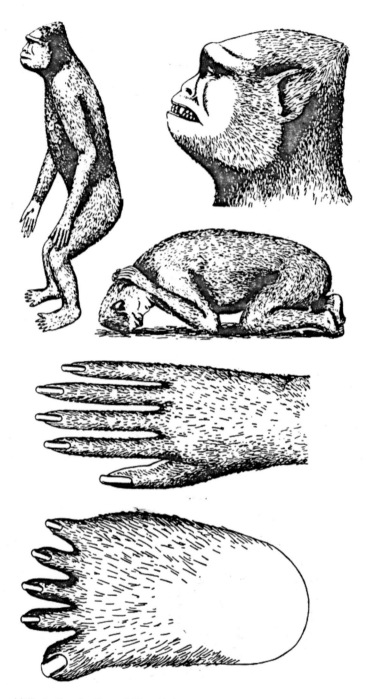

An old illustration for the yeti. The third image is said to show the way in which the creature sleeps. The foot shape in the last image is similar to the second footprint photograph previously shown. *Photo: Public Domain.*

A cavorting yeti shown in an old Tibetan tapestry hanging on a wall in the Thyangboche Monastery. Many such tapestries show yeti-like figures, often in playful or mischievous moods. *Photo: P. Byrne.*

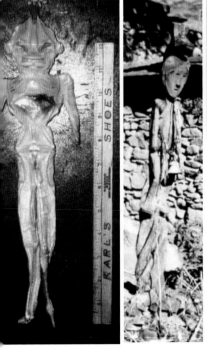

(Above) A completely natural stone formation found in a Himalayan trail by the author. Its similarity to a human footprint is entirely coincidental.

(Far left) Figure given to the author by villagers who stated that it was a mummified infant yeti. It turned out to be a small, turnip-like plant.

(Center) A wooden statue of a shookpa, the Nepalese name for the yeti, in the yard of a Nepalese home in the middle hills. The Sherpa name for the little primates is yeti; the Tibetan name is metah kangmi. *Photos: P. Byrne.*

Gearing up and sorting porter loads for the 1957 (first) reconnaissance. The photograph was taken at the Dharan British Army Gurkha Recruiting Depot on the edge of the foothills in south eastern Nepal. The author is seen with Tom Slick and Dr. N.D. (Andy) Bachkheti, a biologist who came on this first reconnaissance. *Photo: P. Byrne.*

Equipment and food for the 1957 yeti reconnaissance. It was packed in tinfoil-lined, plywood tea chests. Each chest constituted a porter load of about 85 pounds. One of these chests was recently found being used as a table by the headman of an Arun River valley village—54 years later! *Photo: P. Byrne.*

A typical small Nepalese village of about 100 inhabitants in the middle Himalayan ranges.
Photo: P. Byrne.

The author's 1958 yeti expedition team. In the center (back to the white house post and wearing a khaki shirt) is veteran abominable snowman hunter Gerald Russell, of New York. To his left (red and black plaid shirt) the author. Next left (dark green shirt) the author's brother, Bryan. To Bryan's left (red shirt) Gyalzen Norbu, of Darjeeling, the sirdar (foreman) of the Sherpa team. To Gyalzen's left, Chuwang Sherpa, also of Darjeeling, the expedition cook. To Gerald Russell's right (dark shirt and hat) the expedition's Nepalese government liaison officer, Major Pushkar Shumsher Jung Bahadur Rana. Behind him, George Holton, of New York, expedition photographer. Gerald Russell and George Holton are now deceased, and with one exception—Ang Namgyal, of Kathmandu, the expedition's great fifty-miles-a day dak wallah (mail runner) seen fourth from right in front row, with dogs—all of the expedition's loyal, hard working and courageous Sherpa team are no longer with us. Not a few of them, including the expedition's great sirdar, Gyalzen Norbu, died as a result of fatal accidents in the high Himalaya. *Photo: P. Byrne.*

133

Sherpa village of Thammi in the Sola Khumbu district. Behind is the Tesi Lapcha massif and the 18,500-foot pass to the Lang Tang Himal region. *Photo: P. Byrne.*

Sherpa hillmen at Walung village, high above the Arun Khola (Arun River) in the northeastern Himalaya. Three of the men have massive kukri knives tucked into their belts, tools in peacetime, weapons in war. They wear home-made, yak's wool clothing and spend their whole lives shoeless, walking hundreds of miles barefoot over rugged terrain in all kinds of conditions including mud, snow, ice and through frigid Himalayan mountain streams. *Photo: P. Byrne.*

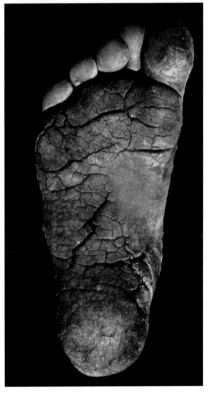

The moonlike surface of the foot of a Nepalese hillman, showing the wear and tear of years of walking barefoot in rugged country. It is interesting and noticeable that in spite of the damage done to the surface of the feet the fine lines of the skin's dermal ridges still exist and can be seen.
Photo: P. Byrne.

The great Sherpa gompa, or monastery, of Thyangboche, in the Sola Khumbu district of northern Nepal. The author's team camped at the monastery many times. Behind it is the magnificent peak of Ama Dablam which rises 23,349 feet. *Photo: P. Byrne.*

(Opposite) Yaks resting in a field in Sola Khumbu. The unknown mountain peak in the background is typical of the hundreds of magnificent 20,000-foot and higher peaks of the great Himalayan range.
Photo: P. Byrne.

Tom Slick's yeti expedition team picks its way through the rugged and often dangerous terrain of the upper Himalaya. *Photo: P. Byrne.*

A Himalayan peak framed in the flaps of a yeti expedition tent. One wakes to splendor.
Photo: P. Byrne.

Himalayan dawn and coffee time. The author at his campsite, 13,500 feet, in a meadow near the Sherpa village of Thammi.
Photo: P. Byrne.

Bryan Byrne and his Sherpa companion, Nima Tenzing, traversing an ice field at 19,000 feet on the approach to the pass of Amphu Lapcha. *Photo: P. Byrne.*

Author with yeti expedition sirdar (Sherpa team foreman) Gyalzen Norbu (red shirt) and two Sherpa porters, emerging from a night spent in a cave. The advantages of living in caves, as opposed to tents, included protection against sub zero, nocturnal, high altitude winds and also the warmth of fires. During all the author's years spent in the high Himalaya, of more than one thousand nights spent at high altitudes, half were in deep rock caves. *Photo: P. Byrne.*

Author assisting heavily laden porters to get across a stream. The high Himalaya is channeled and ravined, north to south, by hundreds of rivers and streams, very few of which have bridges.
Photo: P. Byrne.

Author making use of a handy log to get across a cold, fast flowing high-mountain stream.
Photo: P. Byrne.

The author's brother, Bryan, standing beside a typical upper Himalayan river. Rivers of this magnitude formed impassable barriers for team members and porters. Bypassing them often meant many miles of hard trekking through rough country without any trails. *Photo: P. Byrne.*

A Sherpa-made bridge. For some rivers that must be crossed to allow access to villages, the Sherpa people make bridges of mountain bamboo. The support design is cantilever and the resultant structures are fragile and dangerous—and often nerve-wracking to cross. *Photo: P. Byrne.*

A Nepalese bridge. In the middle Himalaya, Nepalese villagers make crude bridges of bamboo poles. These are tied together with bamboo rope, which is made by manually smashing bamboo poles with a rock into string-like strips.
Photo: P. Byrne.

The author's brother, Bryan, standing on a firm, well-set log bridge across a small rock gorge.
Photo: P. Byrne.

A view of terraced fields. To grow their crops, middle-hill Nepalese people terrace their fields. The terracing makes for attractive patterns, but requires constant attention and maintenance to counteract erosion. *Photo: P. Byrne.*

Bryan Byrne with edible ferns. In the high Himalaya, Peter Byrne's yeti expedition teams used wild, natural foods whenever and wherever these were available. Edible ferns, as seen here, grow in great profusion in the thickly forested 10,000-foot ranges. Other local foods are wild strawberries, wild yellow raspberries, wild rose hips and stinging nettles. There can be little doubt that these plant foods, along with perhaps many others (known and unknown) are familiar to yeti. Essentially, the Himalayan forests are a cornucopia of edible plants and fruits. There is also plenty of both small and large game, and in some areas fish. Add to this an abundance of fresh water and natural cover in the form of dense forests and we have the perfect environment for a yeti. *Photo: P. Byrne.*

A lifelike model of a Loch Ness monster in a natural pool near the town of Drum-nadrochit. *Photo: P. Byrne.*

SECTION THREE

The Loch Ness Monster Mystery

CHAPTER 33

The History of the Beast

One of the first things that should be noted about the Loch Ness monster phenomenon is that it has a long history. And why should we take note of this? We should do so because a background to any mysterious occurrence, in the form of a historical record, strongly supports its authenticity. True, history can be distorted with careless recording, but when it is not, it supplies us with a foundation on which to base the credibility of the occurrence. And credibility goes a long way towards creating a belief in its reality.

The history of the Loch Ness mystery goes all the back to the sixth century, to an account by the venerable Dean of St. Benedict's Abbey at Fort Augustus (western end of the lake). He recorded that one evening, while he was walking in the monastery gardens in the company of his fellow monks, they saw a huge, serpent-like creature rising out of the waters of the lake to pause and stare at them with a frightening look. And, although this part is not recorded, probably sending them all running for their lives. Whatever their reaction, the Dean did record the incident and that record is extant in the annals of the abbey to this day.

From that ancient time to the present day, sightings of strange plesiosaur-like creatures* continue, to where there is a consensus of opinion—even among hard-core scientists—that the deep, cold waters of this ancient lake may well contain a species of aquatic or semi-aquatic creature unknown to science. The result? One of the world's great mysteries, one that, like the bigfoot of the Pacific Northwest and the yeti of the Himalaya, cries out to be explored, investigated and solved.

This section of the *Monster Trilogy,* written by a veteran monster hunter, is a must for anyone challenged by the mystery concealed in the deep, dark waters of one of the largest lakes in Europe—anyone with the curiosity, the courage and the stamina to undertake a personal expedition that might well bring him face to face with one of

*It makes sense that there is more than one Loch Ness Monster, and evidence suggests this.

the most extraordinary animals of our age. A creature which, unlike the marine monstrosities that live in the depths of the ocean (or what one might find on the impossibly distant planets that float in the cosmos light years away from us) *is to be found*. It is like the bigfoot and the yeti—literally within our own backyards and within reach of anyone with the drive, the interest and the determination to get to the bottom of the mystery.

Loch Ness itself is a large, deep, freshwater lake lying in the southern hills of Scotland. It is believed to have been formed by glacial action during an ice age that covered the whole of Scotland and England some 10,000 years ago. When the ice melted, some of the area's animals, like cave bears, had disappeared, but other animals survived. Perhaps the strange creatures sighted in this deep, cold and ancient lake were among the survivors.

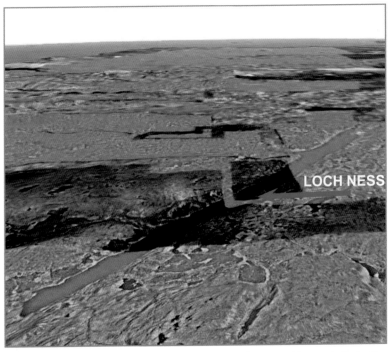

This satellite photograph shows the long, narrow Loch Ness and its proximity to the Atlantic Ocean. *Image from Google Earth; Data S10, NOAA; U.S. Navy; NGA, GEBCO; Image © 2011 Getmapping pic; Image IBCAO.*

THE LOCH NESS AREA

To Dingwall and the North

Caledonian Canal

Weir

Coach for Loch End

Bona

Dores

Temple Pier (Charles W. Wyckoff Station)

To Crannich

Drumnadrochit

Lewiston

Urquhart Castle

Bay

Cobb Memorial

LOCH NESS

Inverfarigaig

Foyers

To Kyle of Lochalsh

Invermoriston

Point Clare

Cherry Island

Knockie Lodge

Horseshoe Scree

Inchnacardoch

Fort Augustus

Glendoe

Abbey

To Spean Bridge and Fort William

Scale: 1 inch = 4.18 miles.

Image: Author's f

150

CHAPTER 34

Early 20th Century Notable Sightings

In the 1930s, Scottish road makers decided to circumvent Loch Ness with a new road. When it was completed, it immediately attracted large numbers of tourists to the area and, rather like the new road in the Bluff Creek area of the Six Rivers National Forest, northern California in 1959, the new lakeside road gave rise to a series of "monster" sightings. The following early reports are noteworthy:

On July 22, 1933 George Spicer and his wife claimed they had a sighting, and it was an unusual one. They said that the creature they saw—huge, serpent like, with a long neck—was actually on the road ahead of them, wriggling across it and heading for the lake. Their story made national news and from that time on sightings began to be made at regular intervals.

On November 12, 1933 Hugh Gray claimed a sighting near Foyers and took a photograph of what he saw. It shows what appeared to be a huge, writhing creature possibly forty feet in length, creating a turbulent disturbance in the water.

In January 1934, another land sighting was claimed, this time by a motorcyclist named Arthur Grant. He stated that at about one o'clock in the morning, as he was riding towards Abriachan (which is on the northeastern side of the lake), he saw a huge creature on the side of the road ahead. The moon was very bright and he could clearly see a long neck topped off by a small, snakelike head. When the creature saw him, it immediately crossed the road in front of him and headed for the lake. Grant got off his motorcycle and ran down to the shore of the lake to get a better look at the creature, but it was gone, the only sign being a heavy disturbance in the water where it had plunged in.

Many other quite credible sightings of the monsters took place through the 1930s and into the period of World War II. During wartime, the lake came under the control of the Royal Navy and was constantly patrolled and observed.

This official Navy "observation period" produced one interesting sighting in May 1943 by Mr. C.B. Farrel, a member of the Royal

Observation Corps. His duty was to watch for enemy aircraft, and one day during that month, while scanning the lake for movement, he saw at a distance of about 250 yards what he described as a huge, dark-colored creature swimming along with some twenty to thirty feet of its body above the surface of the water. Its neck was held vertically and its head was a good five feet above the surface. He could see its eyes, which he said were quite large. Also, he reported that the the creature appeared to have a fin.

During the period from 1934 to the 1980s Alex Campbell, water bailiff at Fort Augustus, was a central figure in the story of the Loch Ness mystery. Campbell claimed to have had sixteen sightings of the creature during this time. I met him in the early 1970s and found him to be both a gentleman and very sincere. For certain, no work on the Loch Ness mystery would be complete without mentioning Alex.

Alex Campbell Photo: P. Byrne.

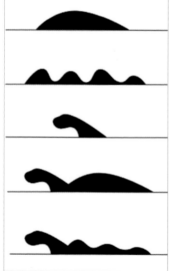

The many possible Nessie sightings have resulted in the chart seen here showing the various shapes of the creature reported by witnesses.
Photo: Public Domain.

CHAPTER **35**

Tim Dinsdale's Nessie Film

To date I have made four visits to Loch Ness. The first was in 1943 when I was in the Royal Air Force and stationed a few miles away from the lake at a place called Invergordon, on the Moray Firth. I was at Invergordon for three months undergoing training as a Catalina flying boat air gunner with the RAF's Air Sea Rescue Service—pending being shipped to the Indian Ocean, where I spent four years. My one visit to the lake was in the company of three friends in a borrowed car, and it was nothing more than a one-day circuit of the lake. We stopped often to look at the loch's cold, grey waters, but did not see any-

Tim Dinsdale Photo: P. Byrne.

thing, and after lunch at a fine inn called the Drumnadrochit Arms, returned to Invergordon. My second trip was quite different. It was many years later and it was in the company of the famous Loch Ness explorer, Tim Dinsdale. We spent fifteen days on the water together and from Tim, a veteran of Nessie hunting, as he called it, I learned a great deal. (Note: Nessie is the creature's nickname, although it applies to probably several of the animals.)

I met Tim Dinsdale through Robert (Bob) Rines, a retired patent attorney who was founder of the esteemed Boston-based Academy of Applied Science, and another great Loch Ness explorer. One day, when I was visiting Bob in Boston to discuss a bigfoot project, I told him that I was on my way to England. He suggested that while there I should try and visit Tim Dinsdale and, if I had the time, to go to Loch Ness with him.

A little later I called Tim from London and introduced myself. I told him that, like him, I was ex-RAF and that I was working with

Bob Rines on a bigfoot project in the United States. Hearing this, he asked me if I had any time to spare during my visit to the UK. When I said yes, he promptly invited me to accompany him to Loch Ness. He said that he would be driving up there and was leaving in a few days.

A couple of days later I took a train from London to Reading, where Tim lived, and after a very pleasant evening and night at his home with he and his wife Wendy, we set out for Scotland by road. We got to the lake two days later and booked in at the Drumnadrochit Arms, near where Tim stored his boat and his Nessie hunting equipment. Three days after leaving Reading we were on the water and looking for Nessie.

Tim's plan of operation on the water was simple: go to selected areas of the lake, preferably where there had been eyewitness reports, and then shut off his little boat's engine and just drift. So this is what we did, cameras at the ready and with endless searching of the waters with high-powered binoculars. A large supply of books helped to fill in the time and each day was broken up by a return to the Drumnadrochit Arms for lunch and a cold beer. After lunch we stayed out on the water until late afternoon and were usually back at the Drumnadrochit Arms by dusk, where we rewarded ourselves for our labors with a dram or two of the inn's best single malt scotch followed by a fine dinner—very often fresh salmon and local farm produce.

Tim operated at the "Ness" only in the summer. He said that winters on the loch were too cold and wet, with storms and turbulent wave conditions that made small boat operations dangerous. So he confined his searching to the warm weather months of Scotland's brief summers, and spent the remainder of the year lecturing on the subject of the monsters and making enough money to support his research.

Simplicity was the watchword of Tim's Loch Ness research, and this was made obvious by the minimal amount of equipment that he owned and used on the water. His boat was quite small—maybe fourteen feet—and propelled by a little five-horsepower outboard motor. The craft had a tiny combination cabin and steering house in which we often huddled together against unexpected squalls of rain. His camera equipment consisted of a single 16mm movie camera (a hand wound Bolex), two 35mm cameras (one with a 300mm lens) and an old pair of naval binoculars (which he purchased in a second-hand store in Reading).

When I joined Tim on the lake for that first time he had been operating there, summer after summer, for almost twenty years, and during that time (he told me with a sigh) he had not seen so much as a ripple on the surface of the cold, dark waters—apart from some bird and fish activity. But he was a man of enormous patience and determination, and he was certain that if he persisted, eventually he would get the pictures and footage that he wanted. To this end he had one advantage over many others—one that fuelled his tenacity—and this was something that had happened many years previously. In April 1960, Tim was at the lake with his family, at a place called Foyers. Scanning the lake for any sign of anything moving, they suddenly noticed a dark object crossing the lake below them, and then turning and cruising parallel to the opposite shore. Tim grabbed his 16mm Bolex movie camera and began shooting. Before the object suddenly dived and disappeared, Tim managed to get some eighty seconds of movie footage.

The footage was sent to the Royal Air Force Joint Reconnaissance Centre for examination and analysis, and was later viewed by another group—the Loch Ness Investigation Bureau. The general consensus of opinions was that the footage was not only genuine, but that it showed what was very possibly one of the legendary creatures swimming half submerged on the surface of the lake. Since then the movie has become something of a "gold standard" for Nessie hunters.

CHAPTER 36

Robert Rines & His Interest in Loch Ness

Some of the most interesting and productive monster investigative work at Loch Ness was carried out by a Boston, Massachusetts-based group called the Academy of Applied Science. The founder of the group was Robert Rines, who passed away in 2009.

I first met Bob in the early 1970s soon after he contacted me about my research on the bigfoot mystery. Our association began with a telephone call. After an in-depth talk about the bigfoot phenomenon, he told me about his great interest in Loch Ness and its giant underwater inhabitants. He then

Robert Rines *Photo: P. Byrne.*

invited me to go to Boston and meet with him and some of his associates of the Academy to discuss the possibility of a task force for an investigation of the bigfoot mystery—his Academy people and resources combined with my background experience and expertise. Part of the agreement that we put together was that I would be invited to go to Loch Ness and take part in some of the Academy's planned work there. In return he and his people would join me from time to time in the Pacific Northwest and assist me with some of the funding (provision of equipment and so forth) that I needed for my work there.

The agreement worked well, and not long afterwards I found myself at Loch Ness working on one of the Academy projects with Bob and getting to know some of the very talented people who made up his crews. I also soon got to know Bob more personally and found him to be (like my yeti expedition sponsor and friend of years before, Tom Slick) a brilliant, talented and fascinating man, with a great zest for life and a broad-minded and positive interest in all three of the great mysteries that so intrigued me—but with a central interest in the strange, unidentified beasts of Loch Ness. Our similar interests and thinking soon gave rise to a friendship that continued to the day he died.

Bob Rine's interests in the Loch Ness monsters began—like that

of many others, including Tim Dinsdale—by something unplanned and unexpected. Bob and his former wife, Carol (who died quite young), were boating on the loch at Urquart Bay in the early 1970s. It was a lovely summer's evening and a sturdy Scottish boatman was rowing them quietly across the bay. Here and there a fish broke the surface—I believe Bob told me that these were Arctic char—but otherwise the surface of the water was like glass, with nothing more than the ripples of their boat marking its surface.

Suddenly, without any warning, the water underneath the boat began to heave and boil. Bob said that it was as though a depth charge had gone off, far down in the murky deeps, generating powerful hydraulics. Next, the water seethed around them, raising the boat a foot above the general loch surface. Bob told me this scared the daylights out of everyone, including the boatman, who promptly spun the boat around and rowed like a madman for the shore. Arriving there, the three of them looked back and saw the waters slowly subsiding to where, after a while, they were as smooth and glassy as before.

The boatman said he had never had an experience like this previously and was quite sure it was one of the loch's monsters passing underneath their boat. He also hinted to Bob that a dram or two of the good stuff, to calm his shaken nerves, "might be a good thing, sir, for which I would be very thankful indeed." Bob agreed to this and he and Carol took the man back to the bar at the Drumnadrochit Arms, where a few drams of the subtle alchemist soon calmed the poor boatman's shaken nerves, and where their extraordinary encounter was the talk of the evening.

This was Bob Rines' introduction to the Loch Ness mystery and from that moment, he told me, there was no turning back. It was a huge and daunting mystery—one that had defeated the best scientific minds in Britain for a hundred years. However, he might well be the man to crack it because not only did he have a scientific background, but now, semi-retired, had time available to get involved and, most important, the financial ability to fund a thorough investigation.

One of the first things that Bob did was to purchase a house at Loch Ness—a fine, four-bedroom residence where later we enjoyed many dinners and spent many comfortable nights. Down on the shore of the lake, right below the house, he had a large boat shed which he would later use for storage of the equipment needed for his Loch Ness monster investigations.

Findings by Robert Rines and His Team

In September 1970, Bob Rines and his crew made their first important discovery on Loch Ness. On the 20th of that month, using sonar, they detected large, unidentified objects intruding into their sonar beam at the same time that fish were seen behaving erratically—as though disturbed by the presence of a predator. The 1970 expedition ended with no other incidents of note, but certainly it raised great expectations.

In early August 1972, Bob and his crew returned to the lake, this time taking with them a stroboscopic underwater camera designed by the great inventor, Harold Edgerton, a personal friend of Bob. In the predawn hours of August 8, 1972, their sonar indicated large moving objects in the presence of which, as before, schools of fish were scattering in what seemed to be alarm. Bob's team was able to home in on one object as it moved within twenty feet of their camera. The camera, located at depth of forty-five feet, was triggered to flash its strobe every fifteen seconds. The resultant extraordinary images showed the hindquarters, a flipper and a section of the tail of a very big creature with rough, textured skin of a dark brown color. The flipper was huge, estimated at six to eight feet in length, indicating that the creature's body was in the region of thirty feet in length.

Various academic institutions examined the findings and all without exception declared that the images were genuine. Scientists at the London-based British Natural History Museum (in spite of their ingrained scientific reticence) also supported the integrity of the images and eventually were moved to state,"the images appeared to show the passage of a large object."

The sonar chart was then analyzed by a number of independent entities, all of whom agreed that it further proved there were indeed large underwater creatures living in Loch Ness—of twenty to thirty feet in length, with bodily segments containing projections that appeared to be humps.

Bob Rines and his crew continued research at Loch Ness for

several more years with modest success. Then in 1975, when again using the Edgerton strobe camera, they made their greatest achievement. This was a set of close-up, underwater photographs that showed in considerable detail the head and part of the body of one of the Loch Ness monsters.

The great value of Bob Rines' enduring research is probably best described by a statement by Dr. George Zug, Curator of Reptiles and Amphibians at the Smithsonian Institution, Washington, DC. Dr. Zug's two-part statement commenced with: "I believe that these data indicate the presence of large animals in Loch Ness." However, he concluded by saying that the findings were insufficient to identify the creatures in question. (Photographs of findings are in the photo album following this section.)

CHAPTER 38

Your Personal Search for Nessie

Getting to Scotland these days to go and hunt the Loch Ness monsters is as easy as getting to the Pacific Northwest to look for bigfoot or getting to Nepal to search for the yeti. There are a dozen international airlines flying in from all over the world with a variety of services that range from super luxury first class to what in the early days of sea travel used to be known as steerage, i.e., knees up on the seat in front and elbows locked with total strangers of forbidding girth. England has a number of international airports, the principal of which are Heathrow and Gatwick. In Scotland, among the best is Edinburgh International Airport from where there is excellent road access to the highlands and the great lake.

Arriving at Edinburgh, one's first move will be to rent a vehicle of some kind. The airport has many car rental agencies and all one has to do is decide on what kind of vehicle is needed. Usually this decision will depend on the condition of the roads that one expects to encounter. In the case of Scotland—and certainly in the general area of Loch Ness—all roads are well paved and provide for comfortable and safe travel. One does not have to think about four-wheel drives, winches, or heavy duty off-road tires, as is the case for bigfoot country in the Pacific Northwest.

If, however, something other than a simple car is envisioned, such as a camper, or a van, then inquiries should be made in advance. Nowadays the internet has a wealth of information on providers of such requirements, including where to find them (other than at airports), vehicle passenger and gear capacity and rental costs. The advantage of using a camper at Loch Ness (as Tim Dinsdale did) is not only that it will provide an uncramped shelter against inclement weather, but parked on the lake shore it will provide a comfortable observation post for searches. I would personally recommend the use of one.

Whatever vehicle you decide on, just remember, if you are American or Canadian, from the minute you get behind the wheel and every day thereafter, that in Great Britain (which includes Scotland) people drive on the left side of the road.

In my present investigation of the bigfoot mystery, the principal method that I apply to my fieldwork is what I call Geo Time Patterns (GTPs). By "fieldwork" (as opposed to general investigative research) I mean physical outdoor searching and observation of selected areas of terrain. GTPs are, simply, patterns of the time and place of credible footprint finds, or credible eyewitness reports, and to some extent, credible historical reports. Behind the design of my GTPs—their very foundation—lies what is essentially the fundamental needs of all living things, which are food, water, shelter and space—needs that apply to everything from an elephant to a mouse, and also to us humans.

Of these vital needs, one of the most important is the need for food on (for many creatures and nearly all wild creatures) a daily basis.

When we humans need food, we simply pick up the phone and order it, or head for the supermarket. In my case, living in the country, we go to the local country store and buy whatever we need (from what is today a vast variety of foods that are both locally produced or imported from all over the world). Wild creatures do not have this kind of access to food, as a result of which, to appease their daily recurring hunger, they must resort to searching for and finding their own natural supply. To do this they may wander great distances and find food wherever and whenever it is available. Or they may simply resort to areas where experience has taught them food is more readily available. For example, a tiger, a big cat that needs a fifty-pound meal of raw meat every five to six days, will spend its whole life within an area of one or two square miles—as long as the area can produce enough food for its needs and also provide for the other survival essentials—water and cover. A grizzly bear, an animal with a huge appetite, will spend its entire life in an area of one square mile, provided the area contains its food and other essential needs as noted.

This is what I believe the bigfoot and yeti do, and indeed it is solidly supported by the figures seen in my Geo Time Patterns. This is also what I believe the Loch Ness monsters may do, with feeding patterns that are seen in the similar locations of the credible sightings. And this is what you should apply to any kind of research at the lake—from moving searches by boat, or from the land, or from stationary observation posts.

To create your Geo Time Pattern for Loch Ness, start by reading

all the books you can find on the subject of its monsters—principally Tim Dinsdale's books—and from these build and develop your GTP data base. You can add to your study the results of inquiries among the people who live around the lake. This latter point is important because books become dated with time, so for current information on credible incidents, local sources will be primary.

As to your "weapons" for Nessie hunting, arm yourself with the very best camcorder, digital camera and telescopic lens you can afford, as well as the very best binoculars. Practice how to use these to the point where you become a "fast gun"—ready to draw and shoot at a moment's notice, because sightings of the lake's mysterious creatures are always, like those of bigfoot and the yeti. fleeting and inevitably conclusive. Five to six seconds is probably all that you are going to get and if you are not ready, then too bad, your opportunity is gone, and more than likely gone for good for the present expedition.

As with bigfoot country, do your reading about appropriate outdoor gear and clothing. Loch Ness and its environs are not as rugged as, say, a camp thirty miles back in the Cascades ranges of Oregon. So survival gear is hardly a requirement, but good warm clothing is appropriate and, for boating out on the lake's long reaches of cold water, and its erratic climate, you need good rain gear.

You can bring your own boat to Loch Ness and use it there as you wish. A permit may be required, and if so, local Information Centres will tell you where to get it. Your boat needs to be something appropriately staunch and suitably matched to the moods of the lake's unpredictable waters which, benign and smiling on a fine summer's morning, can within the space of a few hours turn treacherous and dangerous to small craft.

Local Scottish boatmen have sound boats to rent, of adequate displacement, with broad beams and comfortable cabins for protection against inclement weather. Again, the internet will find them for you—especially Mr. Google—from Invergordon to Inverness, from Fort Augustus to Drumnadrochit.

Good luck and good hunting. Remember: tenacity and patience are the watchwords of your Loch Ness project and, with a little luck, all that you need to come to grips with one of the extraordinary monsters of the dark waters of this ancient Scottish lake—and in doing so make a sterling contribution towards a solution to one of the world's most enduring mysteries.

Loch Ness Monster Photo Album

Provided here are photographs pertaining to my personal investigations at Loch Ness and also those of the Academy of Applied Science (AAS) findings and my interactions with the academy's team.

Tim Dinsdale is seen here in action at Loch Ness. He was one of the most dedicated Loch Ness monster researchers. Between 1960 and 1987, he went solo or led 56 expeditions in search of the creature. He also helped many other researchers. When he first became interested in the possible existence of Nessie, he spent a year analyzing the then-available evidence. In 1960 he borrowed a camera and went on his first expedition. On the last day of this outing, he saw and filmed a large object (creature?) that moved rapidly across, and sometimes below, the surface of the water. He stated that the object appeared larger than any species known to inhabit the loch. He spent the rest of his life in search of further evidence that would compel scientists to take up the search. He influenced many people to consider the reality of the creature through his example, actions, lectures and books. *Photos: P. Byrne.*

On the waters of Loch Ness

John Cobb

having travelled at
206 miles per hour
in an attempt to gain the
World's Water
Speed Record
lost his life
❖ in this bay ❖
Sep: 29th 1952

❖❖

This memorial is erected as a tribute
to the memory of a gallant gentleman
by the people of Glen Urquhart

URRAM DO'N TREUN
AGUS DO'N IRIOSAL

This memorial to John Cobb also contributes to the lore of Loch Ness. Cobb was killed on the loch in 1952 while attempting to break a speed boat record. When such speed trials are held, it is imperative that all normal lake traffic be banned and that the lake water be very still—a condition known as "jelly calm." Such was the case on September 29, 1952, but for some reason Cobb's boat, traveling at 206 miles per hour, took off like a seaplane, somersaulted, and crashed back into the water. Cobb was instantly killed. Film of the tragic accident, taken by a newspaper reporter, was analyzed by Tim Dinsdale. He concluded that the boat seemed to hit a water wake (trail left by something large traveling in water). In his own words, "I had the film step-printed to enable us to show, in slow motion, the very definite line wake and the bouncing of John Cobb's boat when it arrived at the line wake." Dinsdale was convinced that the wake was left by Nessie. In July 2002 Robert Rines and his AAS team, in the process of the search for Nessie using sonar, found what remained of Cobb's boat. *Photo: Wikipedia Creative Commons. Mike Peel, www.mikepeel.net.*

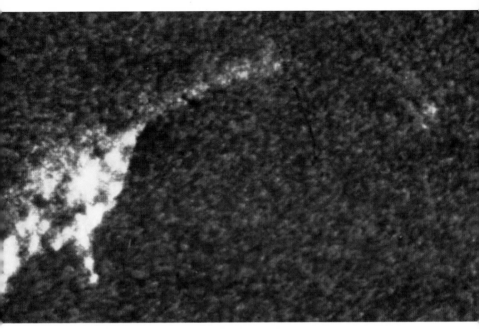

By far the most convincing evidence of Nessie's existence is seen in these two underwater photographs taken on the 1972 expedition headed by Dr. Robert Rines, Academy of Applied Science (AAS), Boston, Massachusetts. In the top photograph we see what appears to be a huge, moving, long-necked creature resembling a plesiosaur. In the bottom photograph, taken at a different place and time, we see what Rines and his AAS team believed to be a huge flipper. It measured approximately four feet in length and, as a motor appendage of that size, indicated a creature of immense girth and weight. *Photos: AAS.*

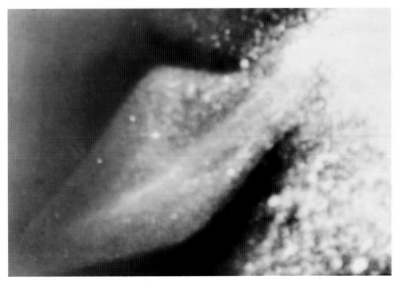

Another photograph taken on the AAS expedition shows what appears to be two large creatures. *Photo: AAS.*

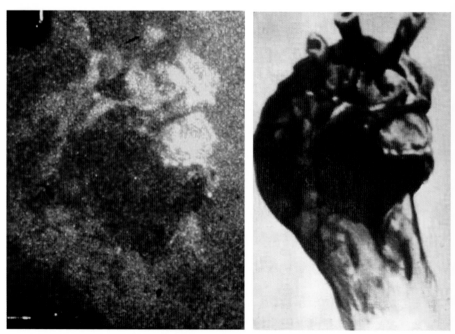

The AAS photograph (left) was speculated to be the head of some sort of creature. An artistic rendering of the head is shown on the right. However, a log was later filmed that skeptics say bore a striking resemblance. *Photos: AAS.*

Dr. Robert Rines, the author's friend and host at Loch Ness, with his sonar scanning submersible. Dr. Rines was an American lawyer, inventor, researcher, and composer with a flair for the unknown. He became well known for his efforts to find the legendary Loch Ness monster.
Photo: P. Byrne.

A close-up of the sonar scanning submersible, which was provided by the Harbor Branch Ocean Discovery Center, Fort Pierce, Florida. This was state-of-the-art equipment at the time and it was greatly anticipated that the Nessie mystery would at last be solved. *Photos: P. Byrne.*

One of the boats used by the AAS team at Loch Ness, the largest and deepest inland expanse of water in Britain. *Photo: P. Byrne.*

A dolphin equipped with a camera for underwater research. This idea was believed to have been originated by Dr. Rines, who confirmed that dolphin training was underway in the United States. He planned to use the process to look for Nessie. It created quite a furor with animal rights groups concerned with the safety of the dolphin and Nessie. The idea was apparently abandonned because of political pressure. *Photo: P. Byrne.*

Dr. Robert Rines beside an AAS sign displaying the name "Charles W. Wyckoff," who was a noted American photographic innovator and photochemist specializing in high-speed photography. It was he who took the famous Nessie photographs on the expedition led by Rines. *Photo: P. Byrne.*

Dr. Robert Rines (right) is seen here with a titled English friend who claimed a Nessie sighting. The two are at the Academy of Applied Science Monument which recognizes Dr. Rines for his discovery of a long-lost World War II plane in Loch Ness.
Photo: P. Byrne.

Dr. Robert Rines with friends and visitors at the Loch Ness Academy of Applied Science Monument. That he was able to find a WW II plane with his sonar equipment would have certainly bolstered his hopes of detecting Nessie. The images he eventually got of what many believe is the creature definitely had the scientific world thinking. After his last expedition in 2008, Rines theorized that the creature may have become extinct, citing the lack of significant sonar readings and a decline in eyewitness accounts. He believed that the creature may have failed to adapt to temperature changes as a result of global warming. He had looked for remains of the creature, again using sonar and an underwater camera, but nothing was found. Dr. Rines died the following year (2009).
Photo: P. Byrne.

Stone engraving found by the author in the ruins of an old monastery near Lough Corrib in western Ireland. The shape of the fin is almost identical to that seen in the famous AAS underwater fin or flipper photograph previously shown.
Photo: P. Byrne.

Another stone "monster" engraving found by the author in the derelict monastery in western Ireland, as detailed in the previous photo caption.
Photo: P. Byrne.

Urquhart Castle on Loch Ness. Although now in ruins, in its day it was one of the largest fortresses of medieval Scotland. The majority of Nessie sightings occur near this castle. *Photo: P. Byrne.*

One of the many fishing or tourist (Nessie hunter's) boats seen daily on Loch Ness. There have been about 4,000 purported sightings of the creature since 1934. Nessie has become a key tourism attraction, bringing an estimated six million pounds (9.5 million US dollars, current rate) a year into the Scottish Highlands. *Photo: P. Byrne.*

Index